VOLTERRA
Theory of Functionals

VITO VOLTERRA
1860-1940

THEORY OF FUNCTIONALS
AND OF
INTEGRAL AND INTEGRO-DIFFERENTIAL EQUATIONS

Vito Volterra

with a Preface by Professor Griffith C. Evans
a Biography of Vito Volterra and a
Bibliography of his Published Works By
Sir Edmund Whittaker

DOVER PUBLICATIONS, INC.
Mineola, New York

DOVER PHOENIX EDITIONS

Copyright

Copyright © 1959 by Dover Publications, Inc.
All rights reserved.

Bibliographical Note

This Dover edition, first published in 1959 and reissued in 2005, is an unabridged republication of the first English translation published by Blackie & Son Limited, Glasgow, in 1930. The Preface by Professor Griffith C. Evans, the Biography and Bibliography by Sir Edmund Whittaker have been added to this edition.

The unusual pagination system of the 1959 Dover edition has been retained here as follows: The front matter is interrupted after page iv to insert the (1959) Preface to the Dover Edition, pages [1] through [40]; the original front matter then resumes on page v to end on page xvi.

International Standard Book Number: 0-486-44284-5

Manufactured in the United States of America
Dover Publications, Inc., 31 East 2nd Street, Mineola, N.Y. 11501

This edition is dedicated to the memory of

VITO VOLTERRA

1860-1940

PREFACE TO THE DOVER EDITION

VITO VOLTERRA often expressed his enthusiasm for the United States, and it is fitting that this book should appear in an American edition. The reprinting contains essentially the prefaces to the previous editions, and is enriched by the inclusion of Sir Edmund Whittaker's Memorial, comprising an account of Volterra's life and an analysis of his scientific contributions. This and the appended bibliography, as well as the suggestive contents of the book itself, should tempt the student to further study in the collected works of the author, now in course of publication in Italy by the Accademia Nazionale dei Lincei. It was my good fortune to study under Professor Volterra from 1910 to 1912.

Volterra first visited America in 1909, on the occasion of the twentieth anniversary celebration of the founding of Clark University, when he delivered the lectures listed as item 136 in the Bibliography. The year 1909 also contributed to the spread of his fame in America through the publication, subsequent to a course of lectures, of Maxime Bôcher's Cambridge Tract "An Introduction to the Study of Integral Equations", which contained a discussion of the algebraic and analytic approaches to the theory of both Volterra and Fredholm.[†] In this little book, Abel, Liouville, Volterra and Fredholm were all represented. A second memorable visit to the United States was the result of President Lovett's invitation to Volterra to participate in the ceremonies of the opening of the Rice Institute in 1912. The lectures which he gave during that festival were printed in Volume III of the beautiful "Book of the Opening", as well as in the Rice Institute Pamphlet [Bibliography 157, 161 and 163]. As a result of his lectures, he became known personally

[†] Cambridge Tracts in Mathematics and Mathematical Physics, London and Edinburgh, 1909.

[1]

PREFACE TO DOVER EDITION

throughout the country — all the way from the Atlantic seaboard to California. He was a member of the American Philosophical Society and the National Academy of Sciences, as well as of the American Mathematical Society and the Econometric Society.

Volterra repeatedly emphasized a certain method of analysis which he described as "that of passing from the finite to the infinite" or "of passing from the discontinuous to the continuous". By this he meant something different from the general process of the passing to the limit that was typical of analysis and the subject of a century and a half of debate. For example, C. S. Peirce was wont to describe the concept of derivative in terms of a passing *through* infinity — however much one should be cautious about dividing by zero. In order to understand the meaning of Volterra's phrases, however, it is necessary to identify oneself with the second half of the nineteenth century, or more precisely, with the student years 1879 to 1880. Riemann had presented the definition of the integral as the limit of a sum in his *Habilitationsschrift* of 1854. Professor U. Dini and the young Volterra were interested in questions concerning integrability, and in 1881 Volterra devoted two papers to this subject (Bibliography, 2 and 3]. In the second of these, *Sui principii del calcolo integrale,* after discussing the upper and lower integrals of a function and conditions of integrability, he goes on to apply the method of upper and lower sums to the existence theorems for systems of ordinary differential equations, as a refinement of the Cauchy treatment of the question, — all this while a student in the Scuola Normale Superiore at Pisa. Evidently Riemann's concept made a great impression on him. Thus in 1887 [Bibliography, 16 and 20] Volterra extended the procedure of Riemann to the definition and determination of the integration of a substitution, with its relation to differentiation and to a linear differential equation. It became his method for the investigation, during the same year, of functionals and functions of curves [Bibliography, 17 and 18] and much of later work, including that on integral equations. A quantity which depends on all the values of a function, or on all the points of a domain is replaced temporarily by one which depends upon a finite number.

Sooner or later a theory begins to generate its own progress and processes. Its form suggests the steps next to come. Thus Volterra puts side by side the two methods for the treatment of integral equations in almost simultaneous papers in the Rendiconti of the Lincei and the

[2]

PREFACE TO DOVER EDITION

Atti of Torino [Bibliography, 56 and 58]. Also in a kind of reverse direction axioms come to be singled out and abstract treatments appear, in which many things are identified, instead of one being used to suggest or generate another. Suffice it to mention Frechet's and Moore's analyses, papers by F. Riesz and the Daniell integral.

This method of Volterra was perhaps for him not only a powerful tool, like the telescope in the hands of Galileo, with which to search in many directions, but also something of an aesthetic principle. A problem in the field was well set if it could be handled in that way. Something similar lay in the dictum, familiar to my generation, that a mathematical problem, for example in partial differential equations, was well set if it corresponded to a physical one; not that all problems in physics, as it turned out, were themselves well set. In any case Volterra was not a man of a single method. His inventions and discoveries enriched many fields in mathematics and its applications. Sir Edmund Whittaker's analysis tells the tale.

Volterra was close to the *Risorgimento*, close to its poets and their national ideals. In 1919 he was prevailed upon, in spite of a modest reluctance, to give a lecture on Carducci. I remember the occasion well because I had the pleasure of translating this, as well as his exposition of functions of composition, *viva voce* to an audience of students. He was also close to the *Rinascimento*, with respect to his sensitivity to art and music and his unlimited scientific curiosity. His devotion to the history of science and his feeling for archaeology were expressed respectively in the treasures of his personal library in Rome and in his collection of antiquities in the villa at Ariccia. He took a most prominent part in the international organization of science and in extending the cultural relations of Italy. His career gives us confidence that the Renaissance ideal of a free and widely ranging knowledge will not vanish, however great the pressure of specialization.

GRIFFITH C. EVANS
Berkeley

February 1958

BIOGRAPHY OF VITO VOLTERRA*

1860-1940

Vito Volterra was born at Ancona on 3 May 1860, the only child of Abramo Volterra and Angelica Almagià. When he was three months old the town was besieged by the Italian army and the infant had a narrow escape from death, his cradle being actually destroyed by a bomb which fell near it.

When he was barely two years old his father died, leaving the mother, now almost penniless, to the care of her brother Alfonso Almagià, an employee of the Banca Nazionale, who took his sister into his house and was like a father to her child. They lived for some time in Terni, then in Turin, and after that in Florence, where Vito passed the greater part of his youth and came to regard himself as a Florentine.

At the age of eleven he began to study Bertrand's *Arithmetic* and Legendre's *Geometry*, and from this time on his inclination to mathematics and physics became very pronounced. At thirteen, after reading Jules Verne's scientific novel *Around the Moon*, he tried to solve the problem of determining the trajectory of a projectile in the combined gravitational field of the earth and moon: this is essentially the 'restricted Problem of Three Bodies', and has been the subject of extensive memoirs by eminent mathematicians both before and after the youthful Volterra's effort: his method was to partition the time into short intervals, in each of which the force could be regarded as constant, so that the trajectory was obtained as a succession of small parabolic arcs. Forty years later, in 1912, he demonstrated this solution in a course of lectures given at the Sorbonne.

---

* This biography of Vito Volterra by Sir Edmund Whittaker appeared in the "Obituary Notices of Fellows of The Royal Society" in 1941. It is reproduced with few omissions and with the permission of The Royal Society. The list of publications which was included has been completed.

BIOGRAPHY

When fourteen he plunged alone, without a teacher, into Joseph Bertrand's *Calcul différentiel:* he does not seem to have had access to any work on the integral calculus at this time, and when he attacked various special problems relating to centres of gravity, he discovered for himself that they could be solved by means of an operation (integration) which was the inverse of differentiation.

His family, whose means were slender, wished him to take up a commercial career; while Vito insisted on his desire to become a man of science. The struggle between vocation and practical necessity became very acute: and the family applied to a distant cousin who had succeeded in the world, to persuade the boy to accept their views.

His interview with his young relative turned out differently from what the family had expected. Impressed by the boy's sincerity, determination and ability, the older man threw his influence on the side of science, and turned the scale. Professor Roiti offered a nomination as assistant in the Physical Laboratory of the University of Florence, though Vito had not yet begun his studies there: it was accepted, and now the die was cast. The young aspirant entered the Faculty of Natural Sciences at Florence, following the courses in Mineralogy and Geology as well as in Mathematics and Physics. In 1878 he proceeded to the University of Pisa, where he attended the lectures of Dini, Betti and Padova: in 1880 he was admitted to the Scuola Normale Superiore, where he remained for three years: and here, while still a student, he wrote his first original papers. Under the influence of Dini he had become interested in the theory of aggregates and the functions of a real variable, and he gave some examples [3][1] which showed the inadequacy, under certain circumstances, of Riemann's theory of integration, and adumbrated the developments made long afterwards by Lebesgue.

In 1882 he graduated Doctor of Physics at Pisa, offering a thesis on hydrodynamics in which certain results, actually found earlier by Stokes, were rediscovered independently. Betti at once nominated him as his assistant. In 1883, when only twenty-three years of age, he was promoted to a full professorship of Mechanics in the University of Pisa, which after the death of Betti was exchanged for the Chair of Mathematical Physics. He now

---

[1] The numbers in square brackets refer to the bibliography at the end of this notice.

BIOGRAPHY

set up house in Pisa with his mother, who up to that time had continued to live with her brother. In 1888 he was elected a non-resident member of the Accademia dei Lincei: in 1892 he became professor of Mechanics in the University of Turin, and in 1900 he was called to the Chair of Mathematical Physics in Rome, as the successor of Eugeno Beltrami. In July of that year he married Virginia Almagià. Six children were born of the union, of whom four now survive. His mother still lived with them, and died at the age of eighty in March 1916.

We must now proceed to an account of Volterra's scientific work. Instead of considering the individual papers one by one in chronological order, we shall group them according to subjects: and shall consider first those relating to functionals.

A *functional* may be introduced as a generalization of the idea of a function y of several independent variables $\phi_1, \phi_2, \ldots, \phi_n$, say $y(\phi_1, \phi_2, \ldots, \phi_n)$. Let us suppose that the set of variables, $\phi_1, \phi_2, \ldots, \phi_n$, is modified from being a finite set to being an enumerably-infinite set and finally to being a continuous set. To represent this analytically, we can regard ϕ_r as a function of its suffix x: then the functional y is a function of all the values that the function $\phi(x)$ takes when x lies in some interval $A \leq x \leq B$. The function $\phi(x)$ is arbitrary, and is, so to speak, the independent variable of which the functional y is a function. This definition may readily be extended to a functional y depending on several functions $\phi_1(x), \phi_2(x), \ldots$ and moreover on certain parameters $t_1, t_2, \ldots, y$ being a function in the ordinary sense of the t's. We may also introduce functions of several variables $\phi(x_1, x_2, \ldots)$ in place of the functions $\phi_r(x)$.

The transition from ordinary functions to functionals corresponds exactly to the transition from the theory of maxima and minima of functions of several variables, to the calculus of variations: and indeed the integrals, which in the calculus of variations are to be made maxima or minima by choosing the functions involved in them in a certain way, constitute a familiar and important example of functionals.

Volterra seems to have conceived the idea of creating a general theory of the functions which depend on a continuous set of values of another function, as early as 1883: but his first published work on the subject [17] did not appear until 1887. The name *functional* was introduced later by Hadamard and has now replaced Volterra's original nomenclature.

BIOGRAPHY

The first step in the theory must evidently be to extend to functionals the well-known fundamental concepts of the theory of functions: the continuity of a functional is first defined, and the derivative and the differential have their analogues. The partial derivative with respect to a particular variable passes into the derivative of the functional γ with respect to ϕ at a certain point of the interval of definition of ϕ, say $x = \xi$. To the total differential, which is a linear form in the differentials of the independent variables with the partial derivatives as coefficients, there corresponds the total variation of the functional, which is an integral over the variation of the independent variables ϕ (ξ) with the derivatives of the functional y with respect to ϕ at the point ξ as coefficients. By repeated application of these operations, higher differential coefficients and higher variations are easily defined. The higher differential coefficients with respect to ϕ at the points $\xi_1, \xi_2, \ldots$ are shown to be symmetrical at these points (corresponding to $\dfrac{\partial^2 f(x,y)}{\partial x \partial y} = \dfrac{\partial^2 f(x,y)}{\partial y \partial x}$). In some cases, however, a slightly generalized definition of the variation is necessary in which, besides the integral, there appears a finite or enumerably-infinite number of terms linear in the variation of ϕ and the derivatives of ϕ at certain exceptional points: the latter is more often than not the case in the Calculus of Variations, where the exceptional points are usually the limits of integration.

In two other notes of the same year, the concept of *functions of a line* was introduced [18]. Let a closed curved line L in space of n dimensions be specified by equations $x_i = \phi_i(t)$ $(i = 1, 2, \ldots, n)$ and suppose that to every such line L there corresponds a definite value of a quantity y: then y is called a *function of the line L*. Obviously y is a functional of the ϕ's: but it is not the most general type of functional, since it is invariant when the ϕ's are replaced by other functions ψ obtained from them by a change of parameter $t = t(u)$, $\psi_i(u) = \phi_i(t)$: y, in fact, depends on the line L but not on its mode of parametric representation. Volterra defined the derivative of a function of a line with respect to the line at a certain point of it, and then defined the variation. He next introduced the idea of a *simple function of a line*: let L_1 and L_2 be two contours having in common an arc which is traversed in opposite directions in the two circuits, and let $L_1 + L_2$ denote the contour obtained by deleting this com-

[8]

mon arc: then a *simple function of a line* is defined to be a function having the property $\phi(L_1 + L_2) = \phi(L_1) + \phi(L_2)$. Volterra established several important theorems regarding these simple functions.

A remarkable application of simple functions of a line is contained in a series of papers, the first of which appeared in the same year [19]. Let L be a line, F and Φ two functions of this line, $L + \Delta L$ a line which is identical with L except in the neighbourhood of a fixed point M, and let $F + \Delta F = F(L + \Delta L)$, $\Phi + \Delta\Phi = \Phi(L + \Delta L)$. Now let the deformation ΔL tend to zero, so ΔF and $\Delta\Phi$ tend to zero. If $\Delta F/\Delta\Phi$ tends to a limit which depends only on M and is independent of the sequence of diminishing deformations ΔL, then the functions F and Φ were said by Volterra to be *connected with each other in Riemann's sense*: he suggested that this corresponds to the relation between two complex variables z and w for which dw/dz is independent of the way in which the limit is approached, depending merely on the value of z. He developed the theory of these functions, showing that it depends on certain partial differential equations, which correspond to the Cauchy-Riemann equations.

In two more notes [21] he gave a theory of the differentiation and integration of connected functions of a line, defining first the 'connexion' between an ordinary function and a function of a line: the limit $d\Phi/dF$ introduced above is shown to be an ordinary function 'connected' with both F and Φ. If f (an ordinary function) and F (a function of a line) are 'connected' and without singularities inside a closed surface σ, then $\int_\sigma fdF = 0$: this corresponds to Cauchy's integral theorem in the theory of analytic functions. Morera's converse of Cauchy's theorem can also be extended to 'connected' functions of a line. Integration and differentiation as introduced in these papers are inverse operations. The theory has some connexions with the theory of analytic functions of two variables.

The whole theory of this generalization of the functions of a complex variable was systematically presented in a considerable memoir [25] in 1889: and at the same time in a series of notes [26, 27, 28] the idea of functions of a line was extended by considering, instead of a line, any sub-space S_r in a space S_n of any number n of dimensions: in particular, the notion of 'conjugate functions' (depending on an S_{r-1} and S_{n-r-1} respectively) was developed. Differential parameters, corresponding to the ∇ and Δ of ordinary theory, were introduced: as in ordinary potential theory, the vanishing of the second differential parameter is a necessary condition for the existence of a conjugate function (the second differential parameter of which also vanishes).

In the following year Volterra showed [30] that by means of his functional calculus the Hamilton-Jacobi theory of the integration of the differential equations of dynamics might be extended to general problems of mathematical physics. The idea was, that whereas the equations of dynamics arise from variational problems relating to simple integrals, the equations of physics arise from variational problems relating to multiple integrals, which must be regarded as functionals of the boundary of the field of integration.

After this, some years elapsed before the work on functionals was continued. In 1892-1894 he published a number of papers [35, 36, 39, 40, 41, 42, 44] on the Partial Differential Equations of Mathematical Physics, especially the equation of cylindrical waves

$$\frac{\partial^2 u}{\partial t^2} = \frac{\partial^2 u}{\partial x^2} + \frac{\partial^2 u}{\partial y^2}.$$

For this equation he obtained an expression of the solution in terms of the initial values of u and $\partial u/\partial t$, which may be regarded as the analogue in two dimensions of the Riemann-Green formula for the propagation of waves in one dimension. He also enquired how Kirchhoff's well-known expression of Huygens' Principle in the wave-theory of light could be extended to space of two dimensions, or of more than three dimensions: and he obtained a formula which gives the value of a cylindrical wave function at a point at the instant t, in terms of the disturbance at points Q of a given curve at the instant $t - PQ/c$ and all previous instants.[2] This formula was used later by Sommerfield in his work on the diffraction of X-rays.

It was after this that Volterra began his celebrated researches in the theory of Integral Equations. He had first met with an integral equation in a paper [11] of

---

[2] These results are brought into relation with more recent work by Hadamard, Marcel Riesz and others, in Baker and Copson, *Huygens' Principle* (Oxford, 1939).

BIOGRAPHY

1884, dealing with the distribution of electric charge on a segment of a sphere: the problem, as he showed, depends on the solution of what would to-day be called an integral equation of the first kind with a symmetric nucleus. It was, however, not until 1896 that he seriously took up work in this field, applying his theory of functionals to what was at that time called the 'inversion of definite integrals', and obtaining results [56, 57, 58] which were greatly admired. In these papers, he regarded the integral equation of the second kind with a variable limit of integration

$$\phi(y) = f(y) + \int_a^y f(x) \, S(x,y) \, dx$$

(where f is the unknown function), which is now generally called an *integral equation of Volterra's type*, as the limiting case of a system of linear algebraic equations; the nth equation of this system contains only the first n unknown quantities, so the system can be solved by recurrence. From the solution of the algebraic system, he passed to the corresponding solution of the integral equation, obtaining a formula

$$f(y) = \phi(y) + \int_a^y \phi(x) \, T(x,y) \, dx$$

where $T(x,y)$ is a function — the *resolvent nucleus*, as it would be called to-day — which can be constructed by a simple process from the given function $S(x,y)$. Unlike later writers — Fredholm, Hilbert, Schmidt — Volterra used the analogy with the linear algebraic equations only heuristically, the final results being proved independently.

In order to deal with integral equations of the first kind

$$\theta(y) - \theta(a) = \int_a^y f(x) \, H(x,y) \, dx$$

(where f is the unknown function), he differentiated this equation with respect to y, thus obtaining an integral equation of the second kind which could be solved by the method already found. A difficulty arises when $H(y,y)$ vanishes or becomes infinite at certain points: he discussed certain types of these 'singular' nuclei, and mastered them.

All these investigations were afterwards extended to the case of multiple integrals, and also to simultaneous

BIOGRAPHY

systems of integral equations, involving several unknown functions. In 1897 he showed [65] that his method is applicable to integral equations with both limits variable: the range of integration actually considered was $ay \leqq x \leqq y$ where $-1 \leqq \alpha < 1$.

In a lecture [68] on the oscillations of liquids under the influence of gravitational forces (the problem of seiches), he advocated the application of infinite determinants to the theory of integral equations — a method which became of great importance later in the work of Fredholm.

In a paper written on the occasion of the centenary of Abel's birth [86], he applied his theory of 'the inversion of a definite integral' to a problem of stability. After pointing out that Abel was the first to solve an integral equation of the first kind of Volterra's type (namely, in the theory of the tautochrone), he discussed the problem of the stability of a fluid mass rotating about one of its principal axes and consisting of concentric, similar, and similarly situated layers: this configuration he proved to be unstable. A simpler proof, independent of the theory of integral equations, was also given: but it was pointed out that this latter proof is less general, as regards the conditions to be satisfied by the function representing the density of the liquid.

Meanwhile, a wide interest in integral equations had been awakened by Fredholm's theory, which was published in Swedish in 1900 and in French (in the *Acta Mathematica*) in 1903. Volterra [101] pointed out the connexion between Fredholm's theory and some problems of his own theory of functionals: the solution of an integral equation is, indeed, a simple case of the solution of a functional equation. In the same paper, and also in the seventh of his Stockholm lectures of the same year, he discussed certain transcendental integral equations originating in the 'Taylor expansion' of the theory of functionals.

To this period of his life belong some celebrated researches in the theory of elasticity, which were important not only on their own account but also because they suggested much of his subsequent work in Pure Mathematics. Perhaps the most notable of these was the theory of what he called *distorsioni*, a term for which Love introduced the English rendering *dislocations*. In elastic solids which occupy a multiply-connected region of space, the displacements may be many-valued functions, corresponding to deformations for which certain fundamental results of the ordinary theory of elasticity are untrue. As a simple

[12]

BIOGRAPHY

example, suppose that a thin slice of material is cut out of an anchor-ring and the new surfaces thus formed are brought into contact and, after having been twisted, joined together. There is then an initial stress even in the absence of all external forces, and there are certain discontinuities in the displacements, although stress and strain are continuous. The joint may in fact be regarded either as a seat of discontinuities in the displacement, or else as a barrier (a branch-cut) for the many-valued functions which represent the components of the displacement.

Such systems had been introduced earlier, notably in Larmor's attempt to explain electrons as places of intrinsic strain in the aether: but Volterra was the first to develop in 1905-1906 [100, 103] a fairly general theory of these 'dislocations'. A comprehensive account of it, with some improvements by E. Cesàro, was published in 1907 [107]. He first determined the many-valued displacement in a multiply-connected region, corresponding to a given one-valued strain: by aid of the formulae thus obtained, he was able to discuss the type of discontinuity at the barrier: this he proved to be of the type of displacement of a *rigid* body. In particular, he discussed in some detail the possible displacements in a hollow cylinder, and also in a system of thin rods: in the former case he was able to compare his deductions with the results of experiment. Once during the war, when he was on a mission to England to discuss scientific questions of common interest to the Allies, he returned after a tiring day to the College where he was a guest, and found that his kind hosts had placed round the walls of his room a number of models of cylinders subjected to Volterra dislocations. He was deeply touched, and often recalled this incident in later life.

His work in elasticity was the origin of his theory of integro-differential equations, i.e. integral equations between the unknown function and its (partial) derivatives. In 1909 [114] he studied a particular type of such equations, and showed that the solution of this integro-differential equation was equivalent to the solution of a simultaneous system consisting of three linear integral equations and a partial differential equation of the second order.

Integro-differential equations occur in various branches of mathematical physics. Thus, for certain substances, the electric or magnetic polarization depends not only on the electromagnetic field at the moment, but also on the history of the electromagnetic state of the matter at all previous instants (*hysteresis*). When the terms corresponding to this physical fact are introduced into the fundamental equations, these become integro-differential equations [115].

A similar situation is found in 'hereditary elasticity' (as Picard called it), to which two of Volterra's notes of the same year are devoted [119, 120]. He assumes 'linear heredity', i.e. that the strain is a linear functional of the stress; in this case the fundamental equations are systems of linear integro-differential equations, and he showed that the strain in a definite interval of time can be determined, provided that the forces in the body and the stress or strain on its surface are known for this time-interval.

In 1910 Volterra introduced into the theory of functionals the fruitful notions of *composition* and of *permutable functions* [127]. The *composition* of two functions $F(x, y)$ and $\Phi(x, y)$ is defined to be the formation of the integral

$$\int F(x, \xi)\Phi(\xi, y)\,d\xi$$

which is denoted by $F*\Phi$, the composition being said to be of the *first kind* if the limits of integration are x and y, and of the *second kind* if the limits are constants a and b: these two cases evidently correspond to Volterra's and Fredholm's integral equations respectively. Two functions F and Φ are said to be *permutable* if their composition is commutative. In the first note, he introduced permutability of the first kind, and in particular investigated functions Φ which are permutable (of the first kind) with a given function F. He transformed the defining equation $F*\phi = \phi*F$ into an integro-differential equation and by solving the latter showed that every function permutable (of the first kind) with F can be represented by a linear aggregate of F and compositions (of the first kind) of F with itself. Since the operation of composition is evidently an extension of the operation of matrix multiplication, to matrices in which the row-number and column-number take a continuous sequence of values, it is obvious that Volterra's result corresponds to the well-known theorem that every matrix which is permutable with a matrix F (whose latent roots are all simple) must be a polynomial in F. The application of Volterra's theorem to the case when F is a constant yielded the important result that the aggregate of all functions permutable of the first kind with a constant is identical with the aggregate of all functions of $(y-x)$.

In two notes of the same year [124, 127] the theory of composition and permutable functions was applied to the theory of integral and integro-differential equations. Consider an algebraic relation $F(z, \zeta) = 0$ between two variables. Let the variables be replaced by two functions f and ϕ and let all multiplications of z with itself or with ζ be replaced by compositions of the corresponding functions. We thus obtain an integral equation between f and ϕ. If it is possible to represent the solution of $F(z, \zeta) = 0$ (where ζ is the unknown) by a power-series in z, then this representation, when z has been replaced by f and the multiplications by compositions, will yield the solution ϕ of the integral equation. Thus the solution of an integral equation has been made to depend on the solution of an algebraic equation. By a similar process the solution of an integro-differential equation may be deduced from the solution of a differential equation.

The results of these notes were applied to the problem of hereditary elasticity in two notes [123, 125] in which Volterra solved the fundamental integro-differential equations for the case of an isotropic sphere. In the second note he also solved a quadratic integral equation.

In 1911 he took up the problem of integro-differential equations with constant limits, i.e. of the Fredholm type [131]. As we have seen, the solution of an integro-differential equation can be reduced to the problem of finding the 'fundamental solution' of a certain associated differential equation. This 'fundamental solution' is represented by a series proceeding according to powers of the parameter which occurs in the integro-differential equation: when this parameter is zero, the integro-differential equation which Volterra was studying, reduced to Laplace's partial differential equation in n variables, and his 'fundamental solution' to the elementary solution of Laplace's equation. In order to solve the integro-differential equation, a series of compositions was derived from the fundamental series: this series of compositions is always convergent in the case of compositions of the first kind, but in the case of integro-differential equations of the Fredholm type, the compositions are of the second kind and the series formed of them do not always converge. Volterra therefore now gave another theory for linear integro-differential equations of Fredholm type, which, however, is applicable only when the coefficients of the integro-differential equations of Fredholm type, which, the second kind). The chief feature of the new theory is that the fundamental solution appears now as the solution of a linear integral equation of Fredholm type (instead of being the solution of a differential equation).

In another note [132] he determined all the functions order two

(i.e. such that $\phi(x, x) = 0$ and $\left(\dfrac{\partial \phi}{\partial y}\right)_{y=x} = 0$)

showing that they depend on an arbitrary function of one variable: he also solved the equation $\psi^*\psi = \phi$ completely, on the assumption that ϕ is a function of order two. In a further note [133] he pointed out that for functions of the special form

$$F(x, y) = \sum_{i, \kappa=1}^{n} a_{ik} f_i(x) \phi_k(y)$$

the question of permutability is identical with the question of commutability of products of the corresponding matrices $(a_{i\kappa})$. Thus from matrix-algebra he was able to find all functions permutable with F (i.e. all matrices permutable with (a_{ik})). The theory was applied to the solution of an integral equation of the nth degree.

Later in 1911 he published [129] a general survey of his method of utilizing the theory of composition and of permutable functions in order to solve integral and integro-differential equations.

In 1912 he gave a more detailed exposition [134] of the theory of integro-differential equations with a variable upper limit and of elliptic type: and (in the same memoir) the theory of 'hereditary elasticity' and of electric and magnetic hysteresis. He also [137] extended his theory of hereditary elasticity by considering vibrations: this led him to integro-differential equations of hyperbolic type, and to some fundamental results regarding vibrations of hereditary type. In many simple cases the 'hereditary' solution may be obtained from the commonly known one by replacing the trigonometric functions of the known solutions by certain transcendental functions which he now defined.

In the following year (1913) he returned once more [145] to integro-differential equations of elliptic type and completed the investigations of 1911 by considering in

BIOGRAPHY

greater detail the case of an odd number of (spatial) dimensions. In a lecture to the Fifth International Congress of Mathematicians at Cambridge [147] he dealt with transcendental integral equations.

In the same year appeared in book form his lectures at Rome on integral and integro-differential equations [143], and his lectures at the Sorbonne on functions of lines [144]; in these a full account is given of the theory of functionals, the basis being always the transition from a finite number of variables to a continuously infinite number. These works did much to make Volterra's ideas widely known. On the invitation of the Berlin Mathematical Society he also delivered a lecture [150] outlining the fundamental notions of the functional calculus and indicating applications to the calculus of variations, integral and integro-differential equations, the theory of quadratic forms in an infinite number of variables, hereditary elasticity, mechanics and electro-magnetism.

In 1914 he published a couple of notes [153, 154] dealing with functional derivative equations, i.e. equations between a functional and its derivative. After studying some simple types of such equations (linear functional derivative equations of the first and second order) he showed that these equations correspond to the differential equations of the ordinary theory. He also discussed a system of integro-differential equations corresponding to a canonical system of equations in dynamics, and obtained the functional derivative equation which corresponds to the Hamilton-Jacobi equation.

In a substantial memoir produced on the eve of the war — in fact, written for the Napier tercentenary in Edinburgh in July 1914 — but not published until 1916 [160, 248], he gave a systematic exposition of his theory of composition of the first kind, introducing some important new ideas. The most notable of these is the 'zero'th compositional power of a function' which plays the part of unity and is essentially identical with Dirac's ∂-function. By the aid of this unit, compositional fractions and hence negative compositional powers are easily defined. Proceeding further in the same direction, a definition of the 'logarithm by composition' is obtained, and the theory of 'functions by composition' developed. In particular, the fundamental notions of the Calculus are extended to the domain of functions by composition, such as: derivative of a function by composition which is proved to be a (new) function by composition, and the definite integral by composition of a function by composition.

BIOGRAPHY

This theory of functions by composition was the subject of three lectures delivered at the Rice Institute, Houston, Texas [172].

Before describing the scientific work of the last twenty-five years of Volterra's life, let us take up again the thread of his personal history. In March 1905 he was created a Senator of the Kingdom of Italy — a great honour for a man still comparatively young — and about this time he was appointed by the Government as Chairman of the Polytechnic School at Turin, and Royal Commissioner. The way was open for him to become a great figure in political and administrative life: but he preferred the career of a pure scientist, and took an active part in public affairs on only two occasions — the Great War of 1914-1918, and the struggle with Fascism.

In July 1914 he was, according to his custom at that time of year, at his country house at Ariccia, when the war broke out. Almost at once his mind was made up that Italy ought to join the Allies: and in concert with D'Annunzio, Bissolati, Barzilai and others, he organized meetings and propaganda which were crowned with success on the 24th of May in the following year, when Italy entered the war. As a Lieutenant in the Corps of Engineers he enlisted in the army, and, although now over fifty-five years of age, joined the Air Force. For more than two years he lived with youthful enthusiasm in the Italian skies, perfecting a new type of airship and studying the possibility of mounting guns on it. At last he inaugurated the system of firing from an airship, in spite of the general opinion that the airship would be set on fire or explode at the first shot. He also published some mathematical works relating to aerial warfare, and experimented with aeroplanes. At the end of these dangerous enterprises he was mentioned in dispatches, and decorated with the War Cross.

Some days after the capitulation of Gorizia he went to this town while it was still under the fire of Austrian guns in order to test the Italian instruments for the location of enemy batteries by sound. At the beginning of 1917 he established in Italy the Office for War Inventions, and became its Chairman, making many journeys to France and England in order to promote scientific and technical collaboration among the Allies. He went to Toulon and Harwich in order to study the submarine war, and in May and October 1917 took part in the London discussions regarding the International Research Committee, to the executive of which he was appointed. He was the first to propose the use of helium as a substitute for hydrogen, and organized its manufacture.

When in 1917 some political parties — especially the Socialists — wanted a separate peace for Italy, he strenuously opposed their proposals: after the disaster of Caporetto, he with Sonnino helped to create the parliamentary *bloc* which was resolved to carry on the war to ultimate victory.

On the conclusion of the Armistice in 1918, Volterra returned to his purely scientific studies and to his teaching work in the University. The most important discoveries of his life after the war were in the field of mathematical biology, and of these we must now give an account.

The title of his discourse [85], at the opening of the academic year following his election to the Roman Chair, shows that in 1901 he was already interested in the biological applications of mathematics. His own researches in this field, stimulated by conversations with the biologist Dr. Umberto D'Ancona of the University of Siena, began at the end of 1925: his first and fundamental memoir [187] (reprinted with modications and additions as [191]: summarized briefly in English [189]: and more fully [200]. Critical summary by J. Pérès, *Rev. gén. Sci. pur. appl.* 38, 285-300, 337-341 (1927)) on the subject appeared in the following year. The theory was developed in several further papers [192, 193, 194, 195] and in the winter 1928-1929 was made the subject of a course of lectures delivered by Volterra at the Institut Henri Poincaré in Paris. These lectures, together with a historical and bibliographical chapter compiled by D'Ancona, were published in 1931 [208].

The entities studied in these investigations were *biological associations*, i.e. systems of animal (or plant) populations of different species, living together in competition or alliance in a common environment: and the theory is concerned with the effects of interaction of these populations with one another and the environment as expressed in their numerical variations.

At the beginning of these researches, Volterra was unaware of similar work already in existence. Some of this[3] referred only to special problems. Other work, such as that of W. R. Thompson on parasitology, though more general in character and allied to Volterra's studies, differed in this respect: that it was necessary to regard generations as distinct and the number of a population as

---
[3] e.g. the analysis of the Ross-Martini malaria equations.

varying discontinuously in time. It will be obvious, however, that there are many problems relating to species in which generations overlap, where the number of a population can more appropriately be regarded, with sufficient approximation, as a continuously varying function of the time: and it is to problems of this kind that Volterra's methods apply. Such problems had been studied to some extent by A. J. Lotka, and some of Volterra's simpler results referring to associations of two species had been anticipated by him.

To understand the ideas on which Volterra's theory is based, we may consider first the by no means trivial case of a single species. Let the number of the population at the time t be $N(t)$. The simplest assumption of constant birth and death rates leads immediately to the 'Malthusian' equation

$$\frac{dN}{dt} = \epsilon N \qquad (1)$$

giving on integration the geometric law of increase

$$N = N_0 e^{\epsilon t}. \qquad (2)$$

Under more general conditions, there will be corrections to be applied to the right-hand side of (1): and it is of course such corrections which falsify the famous predictions of Malthus regarding human populations.

If the environment will support only a limited number of individuals, we must suppose that the 'coefficient of increase' ϵ is no longer constant, but a decreasing function of N. It is simplest to assume that this decrease is linear so that we have the Verhulst equation

$$\frac{dN}{dt} = (\epsilon - \lambda N) N \qquad (3)$$

where ϵ, λ are constants, ϵ being called in Volterra's later papers the *coefficient of auto-increase* of the population, to distinguish it from the whole coefficient of increase $\epsilon - \lambda N$. Integrating (3), we obtain the well-known 'logistic curve'

$$N = \epsilon/(\lambda + ek^{-\epsilon t}) \qquad (4)$$

which is verified observationally in many contexts in biology and economics.

Under more general conditions further corrections must be allowed for; thus in a complicated case we might have to consider an integro-differential equation such as

$$\frac{dN}{dt} = \left\{ \epsilon - \lambda N + \kappa \sin \nu t - \int_0^t N(\tau) f(t - \tau) d\tau \right\} N + \alpha.$$

The total coefficient of increase here consists of ϵ, the coefficient of auto-increase of the population, together with three corrections: — λN representing the Verhulst effect of competition within the species: a periodic term arising perhaps from seasonal variations of the environment and generally producing forced oscillations in N: and an integral representing some delayed effect, such as the intoxication of a closed environment by the accumulation of waste products. The final constant α indicates immigration at a uniform rate.

Volterra's concern, however, is with associations of 2, 3, ... or n species, and hence with 2, 3, ... or n differential equations (or integro-differential equations if delayed effects are envisaged). He investigated in considerable detail the association of two species, one of which feeds on the other: we may call them *predators* and *prey*. For this case we have the Lotka-Volterra equations

$$\text{(prey)} \quad \frac{dN_1}{dt} = (\epsilon_1 - \gamma_1 N_2) N_1 \qquad (5)$$
$$\text{(predators)} \quad \frac{dN_2}{dt} = (-\epsilon_2 + \gamma_2 N_1) N_2$$

By themselves, the prey would increase and the predators die out: hence the coefficients of auto-increase ($\epsilon_1, -\epsilon_2$) are respectively positive and negative. The competition terms, since they depend on encounters, are proportional to $N_1 N_2$ and are positive for the predators, negative for the prey. By integrating these equations, Volterra deduced the existence of periodic fluctuations, whose period is independent of γ_1, γ_2 (so, e.g., it would not be affected by increased protection of the prey): the average values of N_1 and N_2 do not depend on their initial values, and are given by

$$\overline{N}_1 = \epsilon_2/\gamma_2, \quad \overline{N}_2 = \epsilon_1/\gamma_1. \qquad (6)$$

The existence of periodic fluctuations in biological associations was already well known from observation: but ecologists had generally considered that it was necessary to seek an explanation of the fluctuations in some external cause, such as the seasons, or human interference. Partly as a result of the Lotka-Volterra analysis, it is now generally admitted that periodic fluctuations in a constant environment may under some circumstances be sufficiently explained by the mere fact of interaction.

If the populations considered in equations (5) are different species of fish, the effect of fishing (i.e. uniform destruction of both species of fish proportionately to their numbers) would be to increase ϵ_2 and to decrease ϵ_1. From (6) it follows that the mean number of predators ($\bar{N}_2$) is decreased, and that of their prey ($\bar{N}_1$) increased. Similarly, a cessation of fishing, such as occurs on a large scale in time of war, will be to the relative advantage of the predator species. This effect had already been observed by D'Ancona in his statistical study of the Adriatic fisheries over the period 1905-1923. He found, that is to say, a temporary increase in the mean relative frequency of the more voracious kinds of fish, as compared with the fish on which they preyed, during the years 1914-1918.

In the *Leçons sur la Théorie mathématique de la Lutte pour la Vie* [208] a general theory of n species is developed, and the suggestion of a dynamical analogy is introduced by the distinction between what he called *conservative* and *dissipative* associations. Without going into the details of the definition, it may be explained that the association (of one species) represented in equation (1) is conservative, that of equation (3) is dissipative. Equations (5) represent a conservative association: while if we modify them by taking account of competition within each species, just as (3) was obtained by modification of (1), we obtain the equations of a dissipative association, namely,

$$\left. \begin{array}{l} \dfrac{dN_1}{dt} = (\epsilon_1 - \lambda_1 N_1 - \gamma_1 N_2) N_1 \\[1em] \dfrac{dN_2}{dt} = (-\epsilon_2 - \lambda_2 N_2 + \gamma_2 N_1) N_2 \end{array} \right\} \quad (7)$$

The effect of this change is that the fluctuations of N_1 and N_2 are now damped: that is, their amplitudes diminish and the association tends with increasing time to its equilibrium state. This, Volterra showed, is a general property of the associations which he called dissipative, and gives an obvious point to the mechanical analogy. He regarded conservative associations as representing ideal situations not generally attained in nature, and supposed that actual associations are more often of the dissipative type.[4]

Finally, in chapter 4 of his book, Volterra extended the theory of two species to the cases, so important in many biological problems, where delayed effects occur. The two differential equations were now replaced by a pair of integro-differential equations. He solved them by a method of successive approximations, and discussed at length the analogy of this case with his previous studies of hereditary phenomena in elasticity and electromagnetism.

The extension of this study of delayed effects to an association of n species was tackled only in Volterra's last publication on mathematical biology [233], which appeared in 1939. In the intervening years, two very different developments of the theory occurred, in what may be called the 'applied' and 'pure' aspects of the subject.

Firstly [230], Volterra collaborated with D'Ancona in surveying the relevant biological literature for confirmation and further applications. The inherent difficulties of this task arose rather from the immense variety of special conditions occurring in different cases and the correct assessment of these conditions in any particular case, than in the mathematical analysis of the conditions once identified. The fact that the authors were able to find the appropriate mathematical model in many different cases is a vindication of Volterra's technique.

Secondly [220, 221, 222, 223, 224, 225], on the 'pure' side, he extended into the very core of classical dynamics the suggestion of an analogy which has been noticed above. This new development, which he described in a lecture [226] at the University of Geneva on 17 June 1937, may be explained simply by reference again to the predator-prey equations (5). With a change of notation, these can be written

$$N_1 = \left(\epsilon_1 - \frac{N_2}{\alpha}\right) N_1, \quad N_2 = \left(-\epsilon_2 + \frac{N_1}{\beta}\right) N_2.$$

---

[4] G. F. Gause, *The Struggle for Existence* (Baltimore, 1934), described experiments which aimed at reproducing some of Volterra's mathematical models with simple biological models, using yeast cells and different species of protozoa in competition. The attempts were not altogether successful, but it may be noted that the periodic oscillations eventually obtained in a predator-prey experiment were of diminishing amplitude.

We deduce easily

$$\alpha N_1 + \beta N_2 - \alpha \epsilon_1 N_1 + \beta \epsilon_2 N_2 = 0,$$

$$\alpha N_1 + \beta N_2 - \alpha \epsilon_1 \int_0^t N_1 dt + \beta \epsilon_2 \int_0^t N_2 dt = \text{a constant, say H.}$$

(8)

Introducing what he called the *quantities of life* of the two populations,

$$x = \int_0^t N_1 dt, \ y = \int_0^t N_2 dt,$$

and regarding these as analogous to co-ordinates in dynamics, we have in (8) the result which Volterra called the *conservation of demographic energy*, namely,

$$T + V = H,$$

where $T = \alpha x + \beta y$ = actual demographic energy

$V = -\alpha \epsilon_1 x + \beta \epsilon_2 y$ = potential demographic energy.

Moreover, if we introduce the function

$$\Phi = (\alpha x \log x + \beta y \log y) + \tfrac{1}{2}(\dot{x}y - \dot{x}y) - V,$$

it can be shown that the original fluctuation equations (5) are equivalent to the Lagrangian equations of motion

$$\frac{d}{dt}\left(\frac{\partial \Phi}{\partial \dot{x}}\right) - \frac{\partial \Phi}{\partial x} = 0, \quad \frac{d}{dt}\left(\frac{\partial \Phi}{\partial \dot{y}}\right) - \frac{\partial \Phi}{\partial y} = 0.$$

It then follows as in classical dynamics that, as far as the variations of N_1, N_2 are concerned, the properties of the association may be summed up in a variational principle

$$\partial \int_{t_0}^{t} \Phi dt = 0. \tag{9}$$

The analogy has been described so far for the special association (5); but Volterra worked it out generally for a conservative association (and also for certain types of dissipative associations) of n species.

Yet one more analogy was described in a lecture [227, 228] to the Réunion Internationale des Mathématiciens. Generalizing the analysis of 'cessation of fishing' described above by means of equations (5), (6), it referred to the changes in the equilibrium state of a conservative association of n species, caused by variation of their coefficients of auto-increase; and it took the form of some principles of reciprocity not unlike those which appear in the theories of elasticity and electrostatics.

Biologists have been apt to criticize Volterra for preoccupying himself so elaborately with abstract mathematical models based on simplifying assumptions remote from the complexities of nature. Yet this, after all, is the procedure on which the triumphs of physical science have been founded. It would be rash to say whether the analogies with physical science which he unearthed will remain what they appear to be at first, and certainly are, *at least*, a clever and remarkable *tour de force* — or whether they will eventually be seen as the germs of a profound biodynamics, essential to the theoretical and economic biology of the future: what is beyond dispute is that his contributions to pure mathematics will be in demand more and more inescapably as mathematical biology develops.

While the researches which have last been described were Volterra's chief interest during the later years of his life, he still from time to time published contributions to pure analysis.

In 1924 appeared the well-known Volterra-Pérès book [183] on the theory of composition and permutable functions, marking the completion of his work in this field: and a course of lectures on the theory of functionals and of integral and integro-differential equations, given in Madrid in 1925 by invitation of the Faculty of Science of the University, were published in Spanish in 1927 [190] and in an English translation [203] in 1930.

Still more significant from the modern point of view was a work on the general theory of functionals written in collaboration with J. Pérès, of which the first of three projected volumes appeared in 1936 [231]. This first volume contains the general principles of the functional calculus and its applications to the theory of integral equations: the second volume was planned to contain the theories of composition, of permutable functions, of integro-differential equations and functional derivative equations, and of Volterra's generalizations of analytic functions: the third volume would deal with some subsidiary topics and with the applications of the functional calculus. Account was to be taken of the modern theory of functions and of abstract spaces, and the complete work would therefore have been of great importance.

BIOGRAPHY

In 1938 Volterra published, in the Borel series of monographs, a work [232] written in conjunction with B. Hostinský, concerning researches whose origins belong to his earlier period. In 1887-1888 he had written two notes [20, 23] and a substantial memoir [16], dealing with the theory of substitutions or matrices, the infinitesimal operations which can be performed upon these, and their applications to the theory of linear differential equations. He regarded the n^2 elements of a matrix of order n as functions of a variable x which was supposed to be real in these earlier papers, the extension to complex variables being given in a later memoir [75]: and he defined the *derivative* and the *integral* of a matrix with respect to x, showing that these two operations are inverse. The value of these concepts is shown when we consider a system of linear differential equations of the first order

$$\frac{dy_i}{dx} = \sum_{j=1}^{n} a_{ij}(x) y; \quad (i = 1, 2, \ldots, n).$$

Volterra showed that the elements of the 'integral' of the matrix of the $a_{ij}(x)$ yields a fundamental system of solutions of the differential equations. There is no difficulty in extending the formulae so as to include the case of non-homogeneous systems of differential equations, or the case of more than one independent variable. Theories of total differentials of matrices, and of double and curvilinear integrals were then developed: and the transformation of these latter into each other led to the introduction of differential parameters. When the variable x is supposed to be complex, the integral of a matrix along a closed contour can be defined, and a calculus of residues developed, the residues depending on the singularities of the elements of the matrix: in short, the main ideas of the theory of functions of a complex variable can be carried over into matrix theory: and an extension is thus obtained of the well-known results of Fuchs on the expansion of the solutions of a linear differential equation in the neighbourhood of one of its singular points. Volterra then went on to study matrices whose elements are one-valued and regular functions of position on a Riemann surface: these he called *algebraic* matrices, and their integrals he named *Abelian* matrices: the theory is an interesting analogue of the ordinary theory of algebraic functions and their integrals. The later chapters of the 1938 monograph contain many developments and applications of Volterra's original work, due chiefly to Hostinský: one of the most important relates to the solution of the celebrated functional equation which Professor S. Chapman introduced in 1928 (*Proc. Roy. Soc.*, 119) in connexion with a problem in diffusion.

Volterra's scientific activity overflowed in many domains quite outside his more usual fields of research. The student of topology, for instance, who reads Professor Lefschetz's admirable monograph on *Analysis Situs* (Paris, 1924), finds therein the photographs of a number of ingenious models constructed by Volterra in order to show how two manifolds, defined in very different ways, may nevertheless be homeomorphous to each other.

There remains to be told the melancholy story of his later years. In 1922 Fascism seized the reins in Italy. Volterra was one of the very few who recognized from the beginning the danger to freedom of thought, and immediately opposed certain changes in the educational system, which deprived the Italian Middle Schools of their liberty. When the opponents of the Fascist Government in the House of Deputies withdrew altogether from the debates, a small group of Senators, headed by Volterra, Benedetto Croce and Francesco Ruffini, appeared, at great personal risk, at all the Senate meetings and voted steadily in opposition. At that time he was President of the Accademia dei Lincei and generally recognized as the most eminent man of science in Italy.

By 1930 the parliamentary system created by Cavour in the nineteenth century had been completely abolished. Volterra never again entered the Senate House. In 1931, having refused to take the oath of allegiance imposed by the Fascist Government, he was forced to leave the University of Rome, where he had taught for thirty years: and in 1932 he was compelled to resign from all Italian Scientific Academies.[5] From this time forth he lived chiefly abroad, returning occasionally to his country-house in Ariccia. Much of his time was spent in Paris, where he lectured every year at the Institut Henri Poincaré: he also gave lectures in Spain, in Roumania, and in Czechoslovakia. On all these journeys he was accompanied by his wife, who never left him, and learnt typewriting in order to copy his papers for him: he was accustomed to say that the signature 'V. Volterra' in his later works represented not Vito but Virginia Volterra.

---

[5] He was however, on the nomination of Pope Pius XI, a member of the Pontifical Academy of Sciences, and this honour he retained until his death.

BIOGRAPHY

In December 1938 he was affected by phlebitis: the use of his limbs was never recovered, but his intellectual energy was unaffected, and it was after this that his two last papers [233, 234] were published by the Edinburgh Mathematical Society and the Pontifical Academy of Sciences respectively. On the morning of 11 October 1940 he died at his house in Rome. In accordance with his wishes, he was buried in the small cemetery of Ariccia, on a little hill, near the country-house which he loved so much and where he had passed the serenest hours of his noble and active life.

He had received countless honours. He was elected a Foreign Member of our Society in 1910, and had received a similar distinction from almost every National Academy and Mathematical Society in the world; and he was a doctor *honoris causa* of many Universities: in this country, of Cambridge, Oxford and Edinburgh. In his native land he had the Gran Cordone della Corona d'Italia and the Croce di Guerra, and was a Senator of the Kingdom and a Knight of SS. Maurice and Lazarus. In France he was a Grand Officier de la Légion d'Honneur: he had also the orders of Leopold of Belgium, of S. Carlo of Monaco and of the Polar Star of Sweden. On 23 August 1921 he received from King George V the dignity of an honorary K.B.E.: and this was an immense gratification to him, for he was deeply attached to his many friends in this country.

In the words in which his death was announced to his fellowmembers of the Pontifical Academy of Sciences, we believe him *ex hac vita in scientiarum profectum sedulo impensa ad sapientiae aeternitatem transisse.*

E. T. WHITTAKER

LIST OF PUBLICATIONS OF VITO VOLTERRA*

(1) Sul potenziale di un' elissoide eterogenea sopra sè stessa. *Nuovo Cim.*, 9, 221-229 (1881).
(2) Alcune osservazioni sulle funzioni punteggiate discontinue. *G. Mat.*, 19, 76-87 (1881).
(3) Sui principii del calcolo integrale. *G. Mat.*, 19, 333-372 (1881).
(4) Sopra una legge di reciprocità nella distribuzione delle temperatura e delle correnti Galvaniche costanti in un corpo qualunque. *Nuovo Cim.*, 11, 188-192 (1882).
(5) Sopra alcuni problemi di idrodinamica. *Nuovo Cim.*, 12, 65-96 (1882).
(6) Sulla apparenze elettrochimiche alla superficie di un cilindro. *Atti Accad. Torino*, 18, 147-168 (1882).
(7) Sopra alcune condizioni caratteristiche delle funzioni di una variabile complessa. *Ann. Mat. pura appl.* (2), 11, 1-55 (1883).
(8) Sopra alcuni problemi della teoria del potenziale. *Ann. Scu. norm. sup. Pisa* (Tesi di abilitazione.) (1883).
(9) Sull' equilibrio delle superficie flessibili ad inestendibili. *Atti Accad. Lincei*, 8, 214-218 and 244-246 (1884).
(10) Sopra un problema di elettrostatica. *Nuovo Cim.*, 16, 49-57 (1884).
(11) Sopra un problema di elettrostatica. *R. C. Accad. Lincei* (3), 8, 315-318 (1884).
(12) Sulla deformazione delle superficie flessibili ed inestendibili. *R. C. Accad. Lincei* (4), 1, 274-278 (1885).
(13) Integrazione di alcune equazioni differenziali del secondo ordine. *R. C. Accad. Lincei* (4), 1, 303-306 (1885).
(14) Sulle figure elettrochimiche di A. Guebhard. *Atti Accad. Torino*, 18, 329-336 (1883).
(15) Sopra uno proprietà di una classe di funzioni trascendenti. *R. C. Accad. Lincei* (4), 2, 211-214 (1886).
(16) Sui fondamenti della teoria delle equazioni differenziali lineari. Parte prima. *Mem. Soc. ital. Sci. nat.* (3), 6, no. 8, 104 pp. (1887).
(17) Sopra le funzioni che dipendono da altre funzioni. *R. C. Accad. Lincei* (4), 3, 97-105, 141-146 and 153-158 (1887).
(18) Sopra le funzioni dipendenti da linee. *R. C. Accad. Lincei* (4), 3, 225-230 and 274-281 (1887).
(19) Sopra una estensione della teoria di Riemann sulle funzioni di variabili complesse. I. *R. C. Accad. Lincei* (4), 3, 281-287 (1887).

---

* From No. 234 through No. 270 are listed publications not included in Sir Edmund Whittaker's original list.

LIST OF PUBLICATIONS

(20) Sulle equazioni differenziali lineari. *R. C. Accad. Lincei* (4), 3, 393-396 (1887).
(21) Sopra una estensione della teoria di Riemann sulle funzioni di variabili complesse. II, III. *R. C. Accad. Lincei* (4), 4, 107-115 and 196-202 (1888).
(22) Sulle funzioni analitiche polidrome. *R. C. Accad. Lincei* (4), 4, 355-362 (1888).
(23) Sulla teoria delle equazioni differenziali lineari. *R. C. Circ. mat. Palermo*, 2, 69-75 (1888).
(24) Sulla integrazione di una sistema di equazioni differenziali a derivate parziali che si presenta nella teoria delle funzioni conjugate. *R. C. Circ. mat. Palermo*, 3, 260-272 (1889).
(25) Sur une généralisation de la théorie des fonctions d'une variable imaginaire. 1er Mémoire. *Acta Math., Stockh.*, 12, 233-286 (1889).
(26) Delle variabili complesse negli iperspazi. *R. C. Accad. Lincei* (4), 5, 158-165 and 291-299 (1889).
(27) Sulle funzioni conjugate. *R. C. Accad. Lincei* (4), 5, 599-611 (1889).
(28) Sulle funzioni di iperspazi e sui loro parametri differenziali. *R. C. Accad. Lincei* (4), 5, 630-640 (1889).
(29) Sulle equazioni differenziali che provengono da questioni di calcolo delle variazioni. *R. C. Accad. Lincei* (4), 6, 43-54 (1890).
(30) Sopra une estensione della teoria Jacobi-Hamilton del calcolo delle variazioni. *R. C. Accad. Lincei* (4), 6, 127-138 (1890).
(31) Sulle variabili complesse negli iperspazi. *R. C. Accad. Lincei* (4), 6, 241-252 (1890).
(32) Sopra le equazioni di Hertz. *Nuovo Cim.* (3), 29, 53-63 (1891).
(33) Sopra le equazioni fundamentali della elettrodinamica. *R. C. Accad. Lincei* (4), 7, 177-188 (1891) and *Nuovo Cim.* (3), 29, 147-154 (1891).
(34) Enrico Betti. *Nuovo Cim.* (3), 32, 5-7 (1892).
(35) Sulla vibrazioni luminose nei mezzi isotropi. *R. C. Accad. Lincei* (5), 1, 161-170 (1892).
(36) Sulle onde cilindriche nei mezzi isotropi. *R. C. Accad. Lincei* (5), 1, 265-277 (1892).
(37) Sul principio di Huygens. *Nuovo Cim.* (3), 31, 244-255 (1892) and 32, 59-65 (1892).
(38) Sur les vibrations lumineuses dans les milieux biréfringents. *Acta Math. Stockh.*, 16, 153-215 (1892).
(39) Sulle vibrazioni dei corpi elastici. *R. C. Accad. Lincei* (5), 2, 389-397 (1893).
(40) Sulla integrazione delle equazioni differenziali del moto di un corpo elastico isotropo. *R. C. Accad. Lincei* (5), 2, 549-558 (1893).
(41, 42) Sul principio di Huygens. *Nuovo Cim.* (3), 33, 32-36 and 71-77 (1893).
(43) Esercizi di fisica matematica. I. Sulle funzioni potenziali. *Riv. Mat.*, 4, 1-14 (1894).
(44) Sur les vibrations des corps élastiques isotropes. *Acta Math., Stockh.*, 18, 161-232 (1894).
(45) Sulla teoria dei movimenti del polo terrestre. *Astr. Nachr.*, 138, 33-52 (1895).
(46) Sulla teoria dei moti del polo terrestre. *Atti Accad. Torino*, 30, 301-306 (1895).
(47) Sul moto di un sistema nel quale sussistono moti interni stazionarii. *Atti Accad. Torino*, 30, 372-384 (1895).

LIST OF PUBLICATIONS

(48) Sopra un sistema di equazioni differenziali. *Atti Accad. Torino*, 30, 445-454 (1895).
(49) Un teorema sulla rotazione dei corpi e sua applicazione al moto di un sistema nel quale sussistono moti interni stazionarii. *Atti Accad. Torino*, 30, 524-541 (1895).
(50) Sui moti periodici del polo terrestre. *Atti. Accad. Torino*, 30, 547-561 (1895).
(51) Sulla teoria del moti del polo nelle ipotesi della plasticità terrestre. *Atti Accad. Torino*, 30, 729-743 (1895).
(52) Sulla rotazione di un corpo in cui esistono sistemi ciclici. *R. C. Accad. Lincei* (5), 4, 93-97 (1895).
(53) Sul moto di un sistema nel quale sussistono moti interni variabili. *R. C. Accad. Lincei* (5), 4, 107-110 (1895).
(54) Sulle rotazioni permanenti stabili di un sistema in cui sussistono moti interni stazionarii. *Ann. Mat. pura. appl.* (2), 23, 269-285 (1895).
(55) Osservazioni sulla mia nota 'Sui moti periodici del polo terrestre'. *Atti Accad. Torino*, 30, 817-820 (1895).
(56) Sulla inversione degli integrali definiti. *R. C. Accad. Lincei* (5), 5, 177-185 (1896).
(57) Sulla inversione degli integrali multipli. *R. C. Accad. Lincei* (5), 5, 289-300 (1896).
(58) Sull' inversione degli integrali definiti. *Atti Accad. Torino*, 31, 311-323, 400-408, 537-567 and 693-708 (1896).
(59) Lettera al Presidente Brioschi. *R. C. Accad. Lincei* (1896).
(60) Osservazioni sulla nota precedente del Prof. Lauricella e sopra una nota di analogo argomento dell' Ing. Almansi. *Atti Accad. Torino*, 31, 1018-1021 (1896).
(61) Lezioni di meccanica. Prime nozioni di cinematica. Livorno. Giusti 98 pp. (1896).
(62) Sulla rotazione di un corpo in cui esistono sistemi policiclici. *Ann. Mat. pura appl.* (2), 24, 29-58 (1896).
(63) Un teorema sugli integrali multipli. *Atti Accad. Torino*, 32, 859-868 (1897).
(64) Sul principio di Dirichlet. *R. C. Circ. Mat. Palermo*, II, 83-86 (1897).
(65) Sopra alcune questioni di inversione di integrali definiti. *Ann. Mat. pura appl.* (2) 25, 139-178 (1897).
(66) Sulle scarica elettrica nei gas e sopra alcuni fenomeni di elettrolisi. *R. C. Accad. Lincei* (5), 6, 389-401 (1897).
(67) Sulla scarica elettrica nei gas. Editore R. Giusti. Roma 1897.
(68) Sul fenomeno delle seiches. *Nuovo Cim.*, 8, 270-272 (1898).
(69) Sulle funzioni poliarmoniche. *Atti Ist. Veneto* (7), 10, 233-235 (1898).
(70) Sopra una classe di equazioni dinamiche. *Atti Torino*, 33, 451-475 (1898).
(71) Sulla integrazione di una classe di equazioni dinamiche. *Atti Torino*, 33, 542-558 (1898).
(72) Sur la théorie des variations des latitudes. *Acta Math., Stockh.*, 22, 201-296 (1898).
(73) Sur la théorie des variations des latitudes. *Astr. Gesellsch.*, 33, 275-329 (1898).
(74) Sulla scarica elettrica nei gas e sopra alcuni fenomeni di elettrolisi. *Nuovo Cim.* (4), 7, 53-57 (1898).
(75) Sui fondamenti della teoria delle equazioni differenziali lineari. (Parte seconda). *Mem. Soc. Ital. Sci. Nat.* (3), 12, 3-68 (1899).
(76) Sopra una classe di moti permanenti stabili. *Atti Accad. Torino*, 34, 247-255 (1899).

LIST OF PUBLICATIONS

(77) Sul flusso di energia meccanica. *Nuovo Cim.* (4), 10, 337-359 (1899).
(78) Sul flusso di energia meccanica. *Atti Accad. Torino*, 34, 366-375 (1899).
(79) Sopra alcuni applicazioni della rappresentazione analitica delle funzioni del Prof. Mittag-Leffler. *Atti Accad. Torino*, 34, 492-494 (1899).
(80) Sopra alcune applicazioni delle leggi del flusso di energia meccanica nel moto di corpi che si attraggono colla legge di Newton. *Atti Accad. Torino*, 34, 805-817 (1899).
(81) Necrologia del Prof. Eugenio Beltrami. *Annuario Univ. Roma* (1900).
(82) Betti, Brioschi, Casorati, trois analystes italiens et trois manières d'envisager les questions d'analyse. *Congrès intern. Math.*, 43-57 Paris (1900).
(83) Sugli integrali lineari dei moti spontanei a caratteristiche indipendenti. *Atti Torino*, 35, 186-192 (1900).
(84) Sur les équations aux dérivées partielles. *Congr. intern. Math.* 377-378 Paris (1900).
(85) Sui tentativi di applicazione delle matematiche alle scienze biologiche e sociali. Discorso letto il 4 novembre 1901 alla inaugurazione dell' anno scolastico nella R. Università di Roma., 25 pp. (1901).
(86) Sur la stratification d'une masse fluide en équilibre. *Acta Math., Stockh.*, 27, 105-124 (1903).
(87) Sul numero dei componenti indipendenti di un sistema. *R. C. Accad. Lincei* (5), 12, 417-419 (1903).
(88) Commemorazione del Socio Straniero G. G. Stokes. *R. C. Accad. Lincei* (1903).
(89) Sur les équations différentielles du type parabolique. *C. R. Acad. Sci. Paris*, 139, 956-959 (1904).
(90) Relazione sul viaggio compiuto dal Prof. V. Volterra per incarico avuto dalla Commissione nominata per il riordinamento del Politecnico di Torino (1904).
(91) Un teorema sulla teoria della elasticità. *R. C. Accad. Lincei* (5), 14, 127-137 (1905).
(92) Sull' equilibro del corpi elastici più volte connessi. *R. C. Accad. Lincei* (5), 14, 193-202 (1905) and *Nuovo Cim.* (5), 10, 361-385 (1905)..
(93) Sulle distorsioni generate da tagli uniformi. *R. C. Accad. Lincei* (5), 14, 329-342 (1905).
(94) Sulle distorsioni dei solidi elasticiti più volte connessi. *R. C. Accad. Lincei* (5), 14, 351-356 (1905).
(95) Sulle distorsioni dei corpi elastici simmetrici. *R. C. Accad. Lincei* (5), 14, 431-438 (1905).
(96) Contributo allo studio delle distorsioni dei solidi elastici. *R. C. Accad. Lincei* (5), 14, 641-654 (1905).
(97) Note on the application of the method of images to problems of vibrations. *Proc. London Math. Soc.* (2), 2, 327-331 (1905).
(98) Opere del Prof. Alfredo Cornu. *Atti Accad. Torino* (1905).
(99) Fondazione di un Politecnico nella Città di Torino. Discorso pronunciato in Senato, giugno 1906.
(100) Sull' equilibrio dei corpi elastici più volte connessi. *Nuovo Cim.* (5), 10, 361-385 (1905); (5), 11, 5-20, 144-161, 205-221 and 338-347 (1906).
(101) Sur des fonctions qui dépendent d'autres fonctions. *C. R. Acad. Sci. Paris*, 142, 400-409 (1906).

[32]

LIST OF PUBLICATIONS

(102) L'economia matematica ed il nuovo manuale del Prof. Pareto. *G. Economisti* (1906).
(103) Nuovi studii sulle distorsioni dei solidi elastici. *R. C. Accad. Lincei* (5), 15, 519-525 (1906).
(104) Sui tentativi di applicazione delle matematiche alle scienze biologiche e sociali. *Arch. Fisiol.* (1906).
(105) Leçons sur l'intégration des équations différentielles aux dérivées partielles, professées à Stockholm (février-mars 1906) sur l'invitation de S. M. le Roi de Suède. Uppsala, pp. 82 (1906).
(106) Les mathématiques dans les sciences biologiques et sociales. *Revue du mois* (1906).
(107) Sur l'équilibre des corps élastiques multiplement connexes. *Ann. Ecole norm.* (3), 24, 401-517 (1907).
(108) Parole pronunziate alle feste giubilari di Augusto Righi. Bologna (1907).
(109) Il momento scientifico presente e la nuova Società Italiana per il Progresso delle Scienze. *Atti Soc. Ital. Progr. Sc.*, 3-14 (1908).
(110) Parole pronunziate al Congresso della Società Italiana per il Progresso delle Scienze. *Atti Soc. Ital. Progr. Sc.* (1908).
(111) La matematiche in Italia nella seconda metà del secolo XIX. *4th Math. Congr. Rome*, 1, 55-65 (1909).
(112) Sull' applicazione del metodo della imagini alle equazioni di tipo iperbolico. *4th Math. Congr. Rome*, 2, 90-93 (1909).
(113) Giovanni Vailati (Necrologia). *Period. Mat. Inseg. Sec.* (3), 6, 289-292 (1909).
(114) Sulle equazioni integro-differenziali. *R. C. Accad. Lincei* (5), 18, 167-174 (1909).
(115) Sulle equazioni della elettrodinamica. *R. C. Accad. Lincei* (5), 18, 203-211 (1909).
(116) Parole pronunziate al Congresso della Società Italiana per il Progresso delle Scienze. *Atti Soc. Ital. Progr. Sc.* (1909).
(117) Parole del Preside della Facoltà di Scienze. Onoranze al Prof. Cremona. (1909).
(118) Alcune osservazioni sopra proprietà atte ad individuare una funzione. *R. C. Accad. Lincei* (5), 18, 263-266 (1909).
(119) Sulle equazioni integro-differenziali della teoria dell' elasticità. *R. C. Accad. Lincei* (5), 18, 296-301 (1909).
(120) Equazioni integro-differenziali della elasticità nel caso della isotropia. *R. C. Accad. Lincei* (5), 18, 577-586 (1909).
(121) Commemorazione di Valentino Cerruti Roma 1909.
(122) Parole pronunziate avanti al feretro di Stanislao Cannizzaro. *Nuovo Cim.* (5), 19, 387-389 (1910).
(123) Soluzione delle equazioni integro-differenziali dell' elasticità nel caso di una sfera isotropa. *R. C. Accad. Lincei* (5), 19, 107-114 (1910).
(124) Questioni generali sulle equazioni integrali ed integro-differenziali. *R. C. Accad. Lincei* (5), 19, 169-180 (1910).
(125) Deformazione di una sfera elastica, soggetta a date tensioni, nel caso ereditario. *R. C. Accad. Lincei* (5), 19, 239-243 (1910).
(126) Osservazioni sulle equazioni integro-differenziali ed integrali. *R. C. Accad. Lincei* (5), 19, 361-363 (1910).
(127) Sopra le funzioni permutabili. *R. C. Accad. Lincei* (5), 19, 425-437 (1910).
(128) Espacio, tiempo i massa según las ideas modernas. *An. Soc. cient. Argent.*, 70, 223-283 (1911).

LIST OF PUBLICATIONS

(129) Sopra una proprietà generale delle equazioni integrali ed integro-differenziali. *R. C. Accad. Lincei* (5), 20, 79-88 (1911).
(130) Parole del Presidente della Facoltà. Onoranze al Prof. De Helgnero. Roma (1911).
(131) Equazioni integro-differenziali con limiti costanti. *R. C. Accad. Lincei* (5), 20, 95-99 (1911).
(132) Contributo allo studio delle funzioni permutabili. *R. C. Accad. Lincei* (5), 20, 296-304 (1911).
(133) Sopre le funzioni permutabili di 2a specie e le equazioni integrali. *R. C. Accad. Lincei* (5), 20, 521-527 (1911).
(134) Sur les équations intégro-différentielles et leurs applications. *Acta Math., Stockh.*, 35, 295-356 (1912).
(135) Leçons sur l'intégration des équations différentielles aux dérivées partielles. Paris, (1912). (Republication of 105.)
(136) Lectures delivered at the celebration of the twentieth anniversary of the foundation of Clark University, under the auspices of the department of physics. New York and London, (1912).
(137) Vibrazione elastiche nel caso della eredità. *R. C. Accad. Lincei* (5), 21, 3-12 (1912).
(138) Sulle temperature nell' interno delle montagne. *Nuovo Cim.*, 4, 111-126 (1912).
(139) L'évolution des idées fondamentales du calcul infinitesimal. *Revue du mois*, 13, 257-274 (1912).
(140) L'application du calcul aux phénomènes d'hérédité. *Revue du mois*, 13, 556-574 (1912).
(141) Onoranze al Prof. Valentino Cerruti. Roma, (1912).
(142) Henri Poincaré: L'oeuvre mathématique. *Revue du mois*, 15, 129-154 (1913).
(143) Leçons sur les équations intégrales et les équations intégro-différentielles, professées à la Faculté des sciences de Rome en 1910, publiées par M. Tomassetti et F. S. Zarlatti. Collection Borel, pp. 164 (1913).
(144) Leçons sur les fonctions de lignes, professées à la Sorbonne en 1912, recueillies et rédigées par J. Pérès. Collection Borel, pp. 230 (1913).
(145) Sopra le equazioni integro-differenziali aventi i limiti costanti. *R. C. Accad. Lincei* (5), 22, 43-49 (1913).
(146) Sui fenomeni ereditarii. *R. C. Accad. Lincei* (5), 22, 529-539 (1913).
(147) Sopra le equazioni di tipo integrale. *Proc. 5th intern. Congr. Math.*, 1, 403-406 (1913).
(148) Some integral equations. *Bull. Amer. Math. Soc.* (2), 19, 170-171 (1913).
(149) Onoranze al Prof. Dott. G. B. Guccia. *R. C. Circ. Mat. Palermo* (1914).
(150) Les problèmes qui ressortent du concept de fonctions de lignes. *S. B. berl. math. Ges.*, 13, 130-150 (1914).
(151) Osservazioni sui nuclei delle equazioni integrali. *R. C. Accad. Lincei* (5), 23, 266-269 (1914).
(152) Henri Poincaré. *Nouvelle Collection Scientifique*. Paris, 1914.
(153) Sulle equazioni alle derivate funzionali. *R. C. Accad. Lincei* (5), 23, 393-399 (1914).
(154) Equazioni integro-differenziali ed equazioni alle derivate funzionali. *R. C. Accad. Lincei* (5), 23, 551-557 (1914).
(155) Drei Vorlesungen über neuere Fortschritte der mathematischen Physik, gehalten in September 1909 an der Clark Universität. Deutsch von Ernst Lamla. *Arch. Math. Phys., Lpz.* (3), 22, 97-181 (1914). Also Leipzig, pp. 84 (1914).

LIST OF PUBLICATIONS

(156) The theory of permutable functions. Princeton, (1915).
(157) Henri Poincaré. *Rice Inst. Pamphl.*, 1, 133-162 (1915).
(158) Sulle correnti elettriche in una lamina metallica sotto l'azione di un campo magnetico. *Nuovo Cim.* (1914) and *R. C. Accad. Lincei* (5), 24, 220-234, 289-303, 378-390 and 533-543 (1915).
(159) Metodi di calcolo degli elementi di tiro dell' artiglieria aeronautica. *R. C. Inst. Centr. Aero. Roma*, (1916).
(160) Teoria delle potenze dei logaritmi e delle funzioni di composizione. *Mem. Accad. Lincei* (5), 11, 167-269 (1916).
(161) The generalisation of analytic functions. *Rice Inst. Pamphl.*, 4, no. 1, 53-101 (1917).
(162) Relazione sulla missione in Inghilterra ed in Francia. Roma, (1917).
(163) On the theory of waves and Green's method. *Rice Inst. Pamphl.*, 4, no. 1, 102-117 (1917).
(164) Inaugurazione dell' Instituto Centrale di Biologia Marina in Messina. Venezia, (1917).
(165) Pietro Blaserna. *R. C. Senato. Roma*, (1918).
(166) Relazione della conferenza interalleata sulla organizzazione scientifica. *R. C. Accad. Lincei* (1919).
(167) L'entente scientifique. *Nouv. Revue Italie* (1919).
(168) Le congrès de mathématiques de Strasbourg. *Nouv. Revue Italie* (1920).
(169) Sur l'enseignement de la physique mathématique et quelques points d'analyse. *Enseign. Math.*, 21, 200-202 (1920).
(170) Sur l'enseignement de la physique mathématique et quelques points d'analyse (conférence générale). *C. R. congr. intern. math.*, 81-97 (1920).
(171) Commemorazione di Augusto Righi. *Atti Parl. Senato Regno Roma*, (1920).
(172) Functions of composition. Three lectures delivered at the Rice Institute in the autumn of 1919. *Rice Inst. Pamphl.*, 7, no. 4, 181-251 (1920).
(173) Saggi scientifici. Bologna, (1920).
(174) Osservazioni sul metodo di determinare la velocità dei dirigibili. *Rassegna Marittima Aero. Roma*, (1920).
(175) G. Lippmann (Necrologia). *R. C. Accad. Lincei* (5), 30, 388-389 (1921).
(176) A. Ròiti (Necrologia). *R. C. Accad. Lincei* (5), 30, 477 (1921).
(177) The flow of electricity in a magnetic field. Berkeley, pp. 72 (1921).
(178) Funzioni di linee, equazioni integrali e integro-differenziali. *An. Soc. cient. Argent.* (1921).
(179) Les équations aux dérivées fonctionelles et la théorie de la relativité. *Enseign. Math.*, 22, 77-79 (1922).
(180) Mouvement d'une fluide en contact avec un autre en surfaces de discontinuité. *C. R. Acad. Sci. Paris*, 177, 569-571 (1923).
(181) Sur les fonctions permutables. *Bull. Soc. Math. Fr.*, 52, 548-568 (1923).
(182) Da annuncio della morte del Socio Corrado Segre e ne rimpiange la perdita. *R. C. Accad. Lincei* (5), 33, 459-461 (1924).
(183) Leçons sur la composition et les fonctions permutables. Paris, (1924).
(184) Da annuncio della morte del Socio straniero Fusakichi Omori. *R. C. Accad. Lincei* (5), 33, 43 (1924).
(185) Arthur Gordon Webster. Worcester, Mass., 1924.

LIST OF PUBLICATIONS

(186) Commemorazione del Presidente F. d'Ovidio. *R. C. Accad. Lincei* (1925).
(187) Variazioni e fluttuazioni del numero d'individui in specie animali conviventi. *Mem. Accad. Lincei*, 2, 31-113 (1926).
(188) Lois de fluctuation de la population de plusieurs espèces coexistant dans la même milieu. *Association Française Lyon*, 96-98 (1926).
(189) Fluctuations in the abundance of a species considered mathematically. *Nature*, 118, 558-560 (1926).
(190) Teoria de los funcionales y de las ecuaciones integrales e integro-differenciales. Conferencias explicadas en la Facultad de la Ciencias de la Universitad, 1925, redactadas por L. Fantappié. Madrid, pp. 208 (1927).
(191) Variazioni e fluttuazioni in specie animali conviventi. *R. Comit. Talass. Italiano*, Memoria cxxxi, Venezia, (1927).
(192) Sulle fluttuazioni biologiche. *R. C. Accad. Lincei* (6), 5, 3-10 (1927).
(193) Leggi delle fluttuazioni biologiche. *R. C. Accad. Lincei* (6), 5, 61-67 (1927).
(194) Sulla periodicita delle fluttuazioni biologiche. *R. C. Accad. Lincei* (6), 5, 463-470 (1927).
(195) Essai mathématique sur les fluctuations biologiques. *Bull. de la Soc. d'Oceanographie de France*, (1927).
(196) Una teoria matematica sulla lotta per l'esistenza. *Scientia*, 41, 85-102 (1927): French translation, 33-48.
(197) Lois de fluctuations de la population de plusieurs espèces coexistant dans la même milieu. *Ass. Franç. Avanc. Sci.* (1927).
(198) In memoria di H. A. Lorentz. *Nuovo Cim.* (2), 5, 41-43 (1928).
(199) Sur la théorie mathématique des phénomènes héréditaires. *J. Math. pur. appl.* (9), 7, 249-298 (1928).
(200) Variations and fluctuations of the number of individuals in animal species living together. *J. Conseil Int. Explor. Mer*, 3, 1-51 (1928).
(201) In memoria di Traiano Lalesco. *Revista Universitara Bucarest*, 1, 213-215 (1929).
(202) Theory of functionals and of integral and integro-differential equations. Edited by L. Fantappiè. Translated by M. Long. London, (1929).
(203) La teoria dei funzionali applicata ai fenomeni ereditari. *Atti Congresso Bologna*, 1, 215-232 (1929).
(204) Alcune osservazioni sui fenomeni ereditarii. *R. C. Accad. Lincei* (6), 9, 585-595 (1929).
(205) La théorie des fonctionelles appliquée aux phénomènes héréditaries. *Rev. gén. Sci. pur. appl.*, 41, 197-206 (1930).
(206) Sulle fluttuazioni biologiche. *R. C. Semin. mat. fis. Milano*, 3, 154-174 (1930).
(207) Sulle meccanica ereditaria. *R. C. Accad. Lincei* (6), 11, 619-625 (1930).
(208) Leçons sur la théorie mathématique de la lutte pour la vie. Paris, 214 pp. (1931).
(209) (With U. D'ANCONA) La concorrenza vitale tra le specie nell' ambiente marino. 7. *Congr. Int. Agriculture* (1931).
(210) Ricerche matematiche sulle associazioni biologiche. *G. Ist. Ital. Attuari*, 2, 295-355 (1931).
(211) Italian physicists and Faraday's researches. *Nature*, 128, 342-345 (1931).

LIST OF PUBLICATIONS

(212) Le calcul des variations, son évolution et ses progrès, son rôle dans la physique mathématique. (Conf. faites en 1931 à la fac. d. sci. univ. Charles, Praha et à la fac. d. sci. univ. Masaryk, Brno.) Praha and Brno (1932). Czech translation, *Casopis*, 62, 93-116 and 201-227 (1933).
(213) Sur les jets liquides. *J. Math. pur. appl.*, 11, 1-35 (1932).
(214) Équations aux dérivées partielles et théorie des fonctions. *Ann. Inst. H. Poincaré*, 4, 273-352 (1933).
(215) De Moivre's 'Miscellanea analytica'. *Nature*, 132, 898 (1933).
(216) Sur la théorie des ondes liquides et la méthode de Green. *J. Math. pur. appl.* (9), 13, 1-18 (1934).
(217) Représentations des fonctionelles analytiques déduites du théorème de Mittag-Leffler. *J. Math. pur. appl.* (9), 13, 293-316 (1934).
(218) Rémarques sur la note de M. Régnier et Mlle. Lambin. *C. R. Acad. Sci. Paris*, 199, 1684-1686 (1934).
(219) La théorie mathématique de la lutte pour la vie et l'expérience. *Scientia*, 59, 169-174 (1936).
(220) Les équations des fluctuations biologiques et le calcul des variations. *C. R. Acad. Sci. Paris*, 202, 1953-1957 (1936).
(221) Les équations canoniques des fluctuations biologiques. *C. R. Acad. Sci. Paris*, 202, 2023-2026 (1936).
(222) Sur l'intégration des équations des fluctuations biologiques. *C. R. Acad. Sci. Paris*, 202, 2113-2116 (1936).
(223) Le principe de la moindre action en biologie. *C. R. Acad. Sci. Paris*, 203, 417-421 (1936).
(224) Sur la moindre action vitale. *C. R. Acad. Sci. Paris*, 203, 480-481 (1937).
(225) Principes de biologie mathématique. *Acta biotheor.*, 3, 1-36, Leyden (1937).
(226) Applications des mathématiques à la biologie. *Enseign. math.*, 36, 297-330 (1937).
(227) Leggi delle fluttuazione e principii di reciprocità in biologia. *Riv. Biol.*, 22, (1937).
(228) Fluctuations dans la lutte pour la vie, leurs lois fondamentales et le reciprocité. *Conf. Réunion int. mathém.*, Paris, (1938).
(229) Population growth, equilibria and extinction under specified breeding conditions: a development and extension of the theory of the logistic curve. *Human Biology* (Baltimore), 10, 1-11 (1938).
(230) (With U. D'ANCONA) Les associations biologiques au point de vue mathématique. Paris, (1935).
(231) (With J. PÉRÈS) Théorie générale des fonctionelles. Tome 1: Généralités sur les fonctionelles. Théorie des équations intégrales. Collection Borel, Paris, pp. 359 (1936).
(232) (With B. HOSTINSKÝ) Opérations infinitésimales linéaires, applications aux équations différentielles et fonctionelles. Collection Borel, Paris, pp. 238 (1938).
(233) The general equations of biological strife in the case of historical actions. *Proc. Edinburgh Math. Soc.*, 6, 4-10 (1939).
(234) Energia nei fenomeni elastici ereditarii. *Pontif. Acad. Scient. Acta*, 4, 115-130 (1940).
(235) Lezioni di Meccanica Razionale del Prof. Vito Volterra, Professore nella R. Scuola d'Applicazione per gli Ingegneri di Pisa. *Anno Accàdemico*, Autografia Bertini, pp. LXXVIII-408 (1889-90).

LIST OF PUBLICATIONS

(236) Lezioni di Meccanica Razionale del Prof. Vito Volterra, R. Università di Pisa. *Anno Accademico*, Pisa Salvestrini, 1891, pp. (12)-462 (1890-91).

(237) Sinossi delle Lezioni di Meccanica Razionale del Prof. Vito Volterra. *R. Università di Torino*, pp. 402 (1893-94).

(238) Giuseppe Bartolomeo Erba. Notizia biografica. *Annuario della R. Università di Torino*, pp. 145-148 (1896-97).

(239) L'economia matematica ed il nuovo manuale del Prof. Pareto. *Giornale degli economisti* (Roma), s. II, vXXXII, pp. 296-301 (1906).

(240) Discorso pronunciato al Senato sopra la fondazione di un Politecnico nella città di Torino. *Atti parlamentari della Cam?ra dei Senatori*. L Sessione, v. V. pp. 3359-3363 (1904-1906).

(241) Proposta di un' Associazione Italiana per il Progresso delle Scienze. *Atti del Congresso dei Naturalisti Italiani*, Milano (1906).

(242) Onoranze al Prof. Alfonso Sella. *Parole del Preside della Facoltà. di Scienze*. Roma, Bertero, pp. 9-11 (1908).

(243) Giovanni Vailati. *Bollettino della Mathesis*, a. I, pp. 60-63 (1909).

(244) In memoria di Ferdinando De Helguero, Roma, Bertero, pp. 3-5 (1911).

(245) Opere matematiche del Marchese Giulio Carlo dé Toschi di Fagnano Tre volumi, pp. 1232, Milano-Roma-Napoli, Albrighi, Segati (1911-12).

(246) Discorso pronunciato al Senato sopra la istituzione della Scuola di Applicazione per gli Ingegneri presso la R. Università di Pisa. *Atti parlamentari della Camera dei Senatori*. Sessione unica, v. XV. pp. 11675-11678 (1909-1913).

(247) Onoranze a Luciano Orlando, Ruggiero Torelli, Eugenio Elia Levi, Adolfo Viterbi, Professori di Matematica nelle Università italiane caduti in guerra. *Paròle del Preside della Facoltà di Scienze della R. Università di Roma*, Roma, Bertero, pp. 7-8 (1918).

(248) Teoria delle potenze, dei logaritmi e delle funzioni di composizione. *Dedica della memoria all' Università di Edinburgo* (1918).

(249) La terza Conferenza del Consiglio Internazionale di Ricerche tenuta a Bruxelles dal. 18 al 28 Luglio 1919. *L'Intesa Intellettuale*. (Bologna), a. II, pp. 132-150 (1919).

(250) Relazione sull' insegnamento della dinamica nelle scuole industriali. *Rivista d'ottica e meccanica di precisione* (Bologna), v. II, fasc. I, pp. 1-29 (1921).

(251) Commemorazione di Camillo Jordan. *Rendiconti della R. Accademia dei Lincei*, S. V., V. XXXI, pp. 134-135 (1922).

(252) Discorso alla cerimonia inaugurale della prima assemblea generale della Unione Astronomica Internazionale. *Transactions of the International Astronomical Union*, V. I, pp. 127-131. Roma (1922).

(253) Commemorazione di Luigi Pasteur. *Rendiconti della R. Accademia dei Lincei*, S. V., V. XXXII, pp. 403-404 (1923).

(254) Discorso presidenziale del 1924 all' Accademia dei Lincei. *Rendiconti delle sedute solenni della R. Accademia dei Lincei*, V. III, pp. 517-522 (1916-1928).

(255) Discorso presidenziale del 1925 all' Accademia dei Lincei. *Rendiconti delle sedute solenni della R. Accademia dei Lincei*, V. III, pp. 567-573 (1916-1928).

LIST OF PUBLICATIONS

(256) L'ignorance sépare, la science rapproche. Conversation avec M. Vito Volterra, rapportée par Pierre Chanlaine. *La science et la vie.* (Paris) t. XXX, pp. 111-112 (1926).

(257) Rapports et procès-verbaux des réunions du Conseil permanent pour l'exploration de la mer. Discour de clôture. Copenhague, (1927).

(258) Cinquantenaire scientifique de M. Paul Appell. Paris (1927).

(259) La création du Bureau International des Poids et Mesures. Préface. Paris (1927).

(260) Discours prononcé à la Septième Conférence des Poids et Mesures. Comptes rendus des séances. Paris (1927).

(261) Cinquantenaire scientifique de M. Emile Picard. Allocution au nom de l'Academie Royale des Lincei. Paris, Gauthier-Villars (1928).

(262) Erik Ivan Fredholm. *Procès-verbaux des séances du Comité International des Poids et Mésures,* s. II, t. XIII, pp. 277-280 (1929).

(263) I Fisici Italiani e le ricerche di Faraday. *L'Elettrotecnica,* v. XVIII, pp. 806-808 (1931).

(264) Préface à l'ouvrage: Mlle Hélène Freda, Méthode des caracteristiques pour l'intégration des équations aux dérivées partielles linéaires hyperboliques. *Mémorial des sciences mathématiques,* fasc. LXXXIV. Paris, Gauthier-Villars, pp. V-VII (1937).

(265) Préface à l'ouvrage: V. A. Kostitzin, Biologie mathématique. *Collection Armand Colin,* vol. 200, Paris, Colin, pp. 5-7 (1937).

(266) I miei studi più recenti di biologia matematica Gazzetta del Popolo della sera, (1937).

(267) Lois des fluctuations biologiques et leur consequences. *Bulletin de la Société Mathématique de France* (1938).

(268) Conférences sur quelques questions de mécanique et de physique mathématique; I. Rotation des corps dans lesquels existent des mouvements internes. Rédaction par P. Costabel. Paris, Gauthier-Villars, Fasc. IV of the Collection de physique mathématique edited by E. Borel and M. Brillouin (1938).

(269) Rémarques sur l'action toxique du milieu à propos de la Note de M. Régnier et Mlle Lambin. *Comptes rendus de l'Académie des Sciences,* t. CCVII, pp. 1146-1148 (1938).

(270) Calculus of Variations and the Logistic Curve *Human Biology,* v. II, pp. 173-178 (1938).

PREFACE
TO THE ENGLISH EDITION.

The lectures which are published here in English were given by me at the University of Madrid in 1925; they were first published in 1927 by the Faculty of Science of the University, from a text prepared by Professor LUIGI FANTAPPIÈ, and translated into Spanish by Professor OCTAVIO DE TOLEDO.

The book has been improved and corrected and some passages have been completely rewritten. Many additions have been made, due mainly to the large number of papers and memoirs on the subject published in the last few years, and the bibliography has as far as possible been brought up to date and completed by the addition of references to everything published up to the present. These new references (about a hundred in number) and the text have been carefully collated and harmonised. The list of authors contains the names of persons almost all of whom have worked on functionals or closely related subjects.

All this shows that during its forty years of existence the subject has aroused much interest and has penetrated deeply into the various branches of mathematics and their applications. Everything concerning integral, integro-differential, and functional equations, research on functional spaces, the calculus of variations in its broadest sense, questions involving effects of the type described as "hereditary" — all these, in effect, are to-day within the domain of the theory of functionals and make use of general and systematic rules and processes belonging to that theory.

Among the new matter added to the present edition, I wish to mention the results obtained by G. C. EVANS relating to integral equations with a singular kernel and to integro-differential equations, and those obtained by his students, and the references to their applications of the theory to questions of political economy. I also wish to point out that I have discussed much more fully the work of Professor FANTAPPIÈ and the applications he has made of his own results. I have also spoken of the researches of MICHAL on transformations and integral invariants, of MOISIL on the dynamics of continuous media, and of DELSARTE on kinematics, all of which are closely connected with the theory of substitution groups. Lastly, I have summarised the results of my own recent researches into the enefgy equations of hereditary phenomena, and have touched on the studies that have been made of biological fluctuations under conditions involving hereditary effects.

This volume is not the first devoted to the theory of functionals. As long ago as 1913 a course of lectures on the subject given by me at the Sorbonne in 1912 was published in the "Collection Borel"; this series has also contained my lectures on integral and integro-differential equations and on composition and permutable functions, and on the last of these subjects I have also printed the lectures I gave at Princeton University and at the Rice Institute. But it is a pleasure to me to record that the first comprehensive treatise with the title "Functionals" was due to G. C. EVANS, who in 1918 published the volume *Functionals and their Applications: Selected Topics, including Integral Equations*. This consists of five chapters, dealing with functionals, differential operations on functionals, complex functionals and relations of isogeneity, functional equations, and integro-differential equations. Among the last-mentioned EVANS introduces integro-differential equations of the BÔCHER type, i.e. those in which the integrations are along curves or over fields which are arbitrary and variable. The book concludes with the generalisation of the theory of integral equations, which leads Professor EVANS, on the one hand, to MOORE's considerations on general analysis (a subject which has recently been studied by FRÉCHET), and on the other to the theory of permutable functions, which owes much subsequent progress to PÉRÈS. Almost all branches of the whole of the theory are thus contemplated in EVANS's interesting pages.

PAUL LÉVY's treatise, entitled *Leçons d'analyse fonctionnelle*, was published four years later. Here, after a thorough examination of the bases of the functional calculus, the fundamental concepts of functional operations are introduced and subjected to a searching criticism, and his original study of functional derivative equations is then developed in great detail. Professor LÉVY gives an account of GATEAUX's new ideas on multiple integration with an infinite number of variables, on potentials in functional fields of an infinite number of dimensions, and on functional and integro-differential equations; this material was derived from the ideas and rough notes collected and edited by LÉVY himself with devoted care from the papers left by the young mathematician, who died during the war. But a considerable part of the book is justly assigned to the interesting researches carried out by HADAMARD, to whom the theory of functionals owes such important contributions. HADAMARD, in fact (as TONELLI also did later on), places this theory as foundation of the calculus of variations; he discovered the general expression for linear functionals, and, among other applications, gave the functional law of variation of Green's function.

The spread and the development of the theory of functionals owe much to HADAMARD's work. It was he who introduced the happily chosen term "functionals", using this single word to denote what I had introduced and defined many years before, in my first memoir of 1887, by the name of "functions depending on other functions", a term I afterwards shortened into the expression "functions of lines", which is now reserved rather for a more limited class of entities.

It is my earnest wish that these lectures on functionals may encourage students of mathematics to new lines of research and new applications, and may give origin to a systematic and more extensive exposition of all the results here indicated, on the same lines as those followed in the present volume.

The hope of a more intense development and a wider extension of these studies has been revived by the founding of the review *Studia Mathematica*, which has recently been started by Professors STEFAN BANACH and HUGO STEINHAUS. The new review, which is published at Lvov, in Poland, is to be devoted to research in functional analysis.

I cannot end this preface without expressing my warm gratitude to Professors G. C. EVANS, ELENA FREDA, J. PÉRÈS, and G. VACCA, for the valuable suggestions, advice, and help they have given me, and to Miss M. LONG for her care and interest in the by no means easy task of translation.

Rome, June 1930.

VITO VOLTERRA.

PREFACE
TO THE FIRST (SPANISH) EDITION

In April 1925 I had the honour of being invited by the Faculty of Science of the University of Madrid to deliver a course of lectures on "functionals" and the related theories. I wish here to express my sincere thanks to the Faculty, and in particular to its distinguished Dean, Professor OCTAVIO DE TOLEDO.

The present volume consists of the six Madrid lectures, in which I gave a rapid outline of these theories; the arrangement of the subject-matter is indicated by the chapter headings. The lectures were edited and prepared for publication by Dr. LUIGI FANTAPPIÈ, on the basis of my notes and other material, and were translated into Castilian by Professor OCTAVIO DE TOLEDO. To these two expert collaborators, the distinguished professor of the University of Madrid and the younger mathematician from Rome, I have to express my warm gratitude.

From 1887 onwards, the period when I published my earliest researches under the titles "Functions depending on other functions" and "Functions of lines", many articles and several valuable books have been published on this subject; in addition, certain chapters of analysis have undergone important changes in modern treatises as a result of the new concepts introduced along with "functionals". The theories which developed in this way and their applications have taken various directions and differ widely in their nature and scope; while the fundamental principles of functional analysis, exposed as they have been to profound and acute criticism, have gone on gradually developing and acquiring a steadily growing extension.

But a systematic treatise, collecting and co-ordinating the results obtained in the different fields and showing the underlying ideas connecting them, free alike from omissions in the analytical development and from excessive detail — such a work, giving a clear synthetic survey of all that has been done in the domain of "functionals", in my opinion does not yet exist, though many people have asked for it. These lectures certainly do not satisfy this wish, but they indicate the subjects to be discussed and show how these can be divided up into chapters, and may thus provide a framework on which the desired treatise may perhaps in time be constructed. They are accompanied by an extensive bibliography and numerous references to the more important works more or less closely connected with "functionals". I hope, therefore, that students of analysis will find them useful.

VITO VOLTERRA.

CONTENTS

Preface to the Dover Edition - - - - - - [1]
Biography of Vito Volterra - - - - - - [5]
List of Publications of Vito Volterra - - - - [29]

CHAPTER I
FUNCTIONALS.

Section I. Generalities and Definitions.

	Page
§ 1. Generalities	1
§ 2. Definitions	4
§ 3. Functional Fields and Abstract Aggregates	7
§ 4. Continuity	9
§ 5. Linear Functionals	14
§ 6. Functionals of Higher Degree	19
§ 7. Functional Power Series	21

Section II. Operations on Functionals.

§ 1. Differentiation and Derivation	22
§ 2. Derivatives of the Second and Higher Orders	24
§ 3. Extension of Taylor's Theorem	25
§ 4. Other Differentiable Functions	26
§ 5. Calculation of Some Variations	28
§ 6. Integration of Functionals	32
BIBLIOGRAPHY I	35

CHAPTER II.
PROBLEMS OF THE FUNCTIONAL CALCULUS, INTEGRAL EQUATIONS.

§ 1. Generalities	40
§ 2. Definitions	41
§ 3. Volterra's Equations of the Second Kind	43
§ 4. Fredholm's Equations of the Second Kind	48
§ 5. Symmetrical Kernels; Regular Homogeneous Functionals of the Second Degree	50
§ 6. Integral Equations of the First Kind with Constant Limits	51
§ 7. Volterra's Integral Equation of the First Kind	53

XII   CONTENTS

	Page
§ 8. Singular Kernels	54
§ 9. Case in which the Kernel vanishes	60
§ 10. Systems of Integral Equations; Inversion of Multiple Integrals	62
§ 11. General Functional Equations and Implicit Functionals	63
§ 12. Calculation of Solvent Kernels and Approximate Solution	67
BIBLIOGRAPHY II	69

CHAPTER III

GENERALISATION OF ANALYTIC FUNCTIONS.

§ 1. Generalities	74
§ 2. Functions of Lines of the First Degree	77
§ 3. Connectivity of Spaces in Relation to the Polydromy of Functions of Lines of the First Degree	82
§ 4. Case of n-Dimensional Spaces	85
§ 5. Conjugate Functions	86
§ 6. Monogeneity and Isogeneity	90
§ 7. Analytical Continuation in the Space of Monogenic Functions	95
BIBLIOGRAPHY III	97

CHAPTER IV

THEORY OF COMPOSITION AND OF PERMUTABLE FUNCTIONS.

§ 1. Composition and Permutability of the First and Second Kinds	99
§ 2. Powers and Polynomials by Composition	101
§ 3. Series and Functions by Composition of the First Kind	102
§ 4. Integral Addition Theorems	104
§ 5. General Theorem on the Solution of Integral Equations	106
§ 6. Functions Permutable with Unity	108
§ 7. Order of a Function	110
§ 8. Group of the Functions Permutable with a given Function	110
§ 9. Transformations which maintain the Law of Composition, or Pérès's Transformations	114
§ 10. Fundamental Theorem on Permutable Functions	117
§ 11. Powers by Composition with a Fractional Exponent	118
§ 12. Asymptotic Methods	121
§ 13. Fractions by Composition and Powers by Composition with a Negative Exponent	122
§ 14. Logarithms by Composition	124
§ 15. Functions by Composition	129
§ 16. Series by Composition of the Second Kind	132
BIBLIOGRAPHY IV	135

CONTENTS XIII

CHAPTER V

INTEGRO-DIFFERENTIAL EQUATIONS AND FUNCTIONAL DERIVATIVE EQUATIONS.

Section I. Integro-Differential Equations.

		Page
§ 1.	Generalities on Integro-Differential Equations	138
§ 2.	First Examples of Integro-Differential Equations	141
§ 3.	Integro-Differential Equations obtained from the Theory of Permutable Functions of the First Kind	143
§ 4.	Integro-Differential Equations obtained from the Theory of Permutable Functions of the Second Kind	145
§ 5.	Preliminary Considerations on Hereditary Questions. Integro-Differential Equations for the Elastic Torsion of a Wire	147
§ 6.	Integro-Differential Equations of Elliptic Type	149
§ 7.	Integro-Differential Equations of Hyperbolic and Parabolic Type	154

Section II. Functional Derivative Equations.

§ 1.	Extension of the Jacobi-Hamilton Theory	155
§ 2.	Exact total Differential of a Functional	156
§ 3.	Extension of Stokes's Theorem	158
§ 4.	Extension of Euler's Theorem	159
§ 5.	Relations between Linear Functional Derivative Equations and Integro-Differential Equations	161
§ 6.	Equations of Canonical Type	163
§ 7.	Functional Derivative Equations of the Second Order	164
§ 8.	A Functional Derivative Equation of Green's Function	165
	BIBLIOGRAPHY V	167

CHAPTER VI

APPLICATIONS. OTHER DIRECTIONS OF THE THEORY OF FUNCTIONALS.

Section I.

§ 1.	Calculus of Variations	172
§ 2.	Integral Equations	174
§ 3.	Researches of Hilbert, Schmidt, Goursat, Vitali, etc.	175
§ 4.	Seiches	177
§ 5.	Vibrations of Membranes	178
§ 6.	New Researches on Seiches	179
§ 7.	Some Researches of G. C. Evans and of his School	180

Section II.

§ 1.	Permutable Functions; Functional Groups	182
§ 2.	Integral Addition Theorems	182
§ 3.	Functional Transformations. Functional Invariants	183

Section III.

	Page
§ 1. Functional Dynamics	184
§ 2. Functional Rotations	185
§ 3. The Strain-Energy Functional in the Theory of Elasticity	187

Section IV.

§ 1. General Laws of Heredity	188
§ 2. Hereditary Dynamics	191
§ 3. Hereditary Elasticity	192
§ 4. Hereditary Electromagnetism	194
§ 5. Energy Equations	196

Section V.

§ 1. Functional Operators	200
§ 2. Analytic Functionals	201
§ 3. Abstract Spaces	204
§ 4. Various Applications	206
BIBLIOGRAPHY VI	208
INDEX OF AUTHORS	219
INDEX OF SUBJECTS	223

VOLTERRA
Theory of Functionals

CHAPTER I

FUNCTIONALS

Section I. Generalities and Definitions.

§ 1.

1. Before defining the quantities which we propose to study further on, we shall examine a simple problem of maxima and minima which shows how we pass naturally from the consideration of functions of a finite number of variables to quantities which depend on an infinite number of variables, i.e. on the infinite number of values assumed by an arbitrary function $x(t)$ corresponding to the values of t within an interval (a, b).

2. Let us then consider the product of two numbers x, y, whose sum is constant; it is known that this product is a *maximum* when the two numbers are equal. Translating this fact into geometrical language, we can say that among all the rectangles which have the sum of two consecutive sides constant, and therefore have a *constant perimeter*, the square is the one of *maximum area*. Passing to the more general problem of determining, among all plane polygons of n sides whose perimeter is constant, the one of maximum area, it will be seen that the solution is given by the regular polygon of n sides. If we denote the $2n$ coordinates of the n vertices of the polygon by $(x_1, y_1), (x_2, y_2), \ldots (x_n, y_n)$, the quantity to be made a maximum is the area A, given by the expression

$$A = \frac{1}{2} \sum_{1}^{n}{}_i \begin{vmatrix} x_i & x_{i+1} \\ y_i & y_{i+1} \end{vmatrix} = \frac{1}{2} \sum_{1}^{n}{}_i (x_i \Delta y_i - y_i \Delta x_i), \qquad (1)$$

a function, in this case, of the $2n$ variables x_i, y_i, with

$$\sum_{1\ i}^{n}\sqrt{\Delta x_i^2 + \Delta y_i^2} = \text{constant}, \tag{1'}$$

and in which

$$x_{n+1} = x_1, \qquad y_{n+1} = y_1,$$
$$\Delta x_i = x_{i+1} - x_i, \qquad \Delta y_i = y_{i+1} - y_i.$$

But if we pass from the case of a polygon of constant perimeter to the much more general problem of determining, among all the closed curves C of a given length l, the one that includes the maximum area A (the circle), the quantity A that we have to consider is given by the formula

$$A = \frac{1}{2}\int_C (x\,dy - y\,dx), \tag{2}$$

with

$$\int_C \sqrt{dx^2 + dy^2} = l = \text{constant}, \tag{2'}$$

and thus no longer depends on *a finite number of variables, but on the values of the coordinates of all the infinite number of points of the line C*. Instead of a problem of the ordinary differential calculus on maxima and minima, in which we have to determine a certain *finite* number of unknown quantities, we have a problem of the calculus of variations, in which the unknown is a *function* (its values for all the points of the interval considered) or a *curve* (the coordinates of all its points).

3. In the case of the polygon of n sides it was sufficient to find the quantities (x_1, y_1), (x_2, y_2), ... (x_n, y_n), depending on the index i, discontinuous and variable for *integral* values from 1 to n, in order to determine the polygon itself. In the case of any closed curve C, however, it will be necessary, in order to determine it, to give the coordinates x and y of all its points

$$x = x(t), \qquad y = y(t),$$
$$x(b) = x(a), \qquad y(b) = y(a),$$

in which now the quantities x, y depend on the parameter t which *varies continuously* in a certain interval (a, b); this *continuous parameter t* takes the place of the *discontinuous index i*. Corresponding to the formulae (1) and (1'), containing *sums with respect to the index i*, we get instead the formulae (2) and

I GENERALITIES AND DEFINITIONS 3

(2'), containing *integrals* in which we may *take the parameter t as the variable of integration*.

The path followed in this particular example to pass from a problem with a finite number of unknowns to a problem in which the unknown is a function (a problem with an infinite number of unknowns) has the character of a general method. In many other cases in which this eventuality arises, it will in fact be sufficient to *replace a discontinuous index i by a continuous index, or parameter, t, and the sum with respect to this index i by the integral with respect to the variable of integration t*.

4. There are however problems in which the unknown is a function, but which have a different character. Thus, for instance, if we wish to determine a curve, $y = y(x)$, whose subtangent is constant $(= a)$, we shall get the differential equation

$$a \frac{dy}{dx} = y,$$

and in this equation there is no integral with respect to x; all the infinite number of values of the unknown y corresponding to the infinite number of values of x do not simultaneously enter into consideration; we can say rather that this equation establishes a relation solely between *a value of y and the value it has at an infinitely near point*, determined by $\frac{dy}{dx}$. More generally, all the problems that give rise to an ordinary differential equation

$$f(x, y, y', \ldots y^{(n)}) = 0$$

which establishes a relation only between the values taken by y at a point x and at n infinitely near points, are of this type.

5. A substantially different problem, and related to those considered earlier, is that of determining a plane curve (of equation $y = y(x)$) passing through two given points in the plane (of coordinates x_0, $y_0 = y(x_0)$; x_1, $y_1 = y(x_1)$), such that when it rotates round a line fixed in the plane, e.g. the x-axis, it generates the surface of minimum area. In this case, the area A, the quantity to be made a minimum, is given by the expression

$$A = 2\pi \int_{x_0}^{x_1} y \sqrt{1 + y'^2}\, dx,$$

4   THEORY OF FUNCTIONALS   CHAP.

and therefore depends on all the values of y in the interval (x_0, x_1). In general, the problems treated in the calculus of variations all have this character.

§ 2.

6. From these particular examples we can reach a general definition which shall include the various cases so far considered. We shall therefore say that a quantity z is *a functional of the function $x(t)$ in the interval (a, b) when it depends on all the values taken by $x(t)$ when t varies in the interval (a, b)*; or, alternatively, *when a law is given by which to every function $x(t)$ defined within (a, b) (the independent variable within a certain functional field) there can be made to correspond one and only one quantity z, perfectly determined, and we shall write*[1]

$$z = \mathrm{F} \left\| \left[x \overset{b}{\underset{a}{(t)}} \right] \right\|.$$

This definition of a *functional* recalls especially the ordinary general definition of a *function* given by Dirichlet.

If other variables $\alpha, \beta, \ldots$ also occur in the function $x(t)$ (the independent variable), we shall write

$$z = \mathrm{F} \left[x \overset{b}{\underset{a}{(t)}}, \alpha, \beta, \ldots \right] = z(\alpha, \beta, \ldots)$$

to indicate that the functional operator F, which when applied to the variable function x gives the quantity z, is to be applied considering x as a function of t alone, i.e. supposing that in x the quantities $\alpha, \beta, \ldots$ are constant; z will then be an ordinary

---

[1] The notation constantly used from my first paper of 1887 has always been

$$\mathrm{F} \left\| \left[x \overset{b}{\underset{a}{(t)}} \right] \right\|$$

but for the sake of simplicity, when there is no ambiguity, we can use one of the following:

$$\mathrm{F} \,|\, [x(t)] \,| \qquad \text{or} \qquad \mathrm{F} \left[x \overset{b}{\underset{a}{(t)}} \right]$$

or also

$$\mathrm{F}\,[x(t)].$$

This last notation has been used in this edition.

I GENERALITIES AND DEFINITIONS 5

function of $\alpha, \beta, \ldots$, and it is to be noted that it will *not* be a function of t as x was.

The functional z may also contain certain parameters $\lambda, \mu, \ldots$

$$z = \mathrm{F}\left[x\overset{b}{\underset{a}{(t)}}; \lambda, \mu, \ldots\right];$$

in this case the quantity z, for every system of values of the parameters, will be *a functional of* $x(t)$, in the sense stated above, while it will be *an ordinary function* $z(\lambda, \mu, \ldots)$ of the parameters $\lambda, \mu, \ldots$ when $x(t)$ is fixed. We can also say that in this case, corresponding to every function $x(t)$, defined within a certain field of variation, the functional operator F determines another function

$$z(\lambda, \mu, \ldots) = \mathrm{F}\left[x\overset{b}{\underset{a}{(t)}}; \lambda, \mu, \ldots\right].\,(^1)$$

7. This notion of a *functional* can be extended immediately to the quantities z which depend on all the values taken not by one but by several functions, e.g. by two, $x = x(t)$, $y = y(u)$, in the intervals (a, b) and (c, d) respectively:

$$z = \mathrm{F}\left[x\overset{b}{\underset{a}{(t)}}, y\overset{d}{\underset{c}{(u)}}\right].$$

The consideration of these functionals, however, presents very little interest, as they can be immediately reduced to those *depending on a single variable function*. In fact, let $a'b' = a'c' + c'b'$ be an interval of variation of a parameter v, of amplitude equal to the sum of the two intervals (a, b) and (c, d); and let c' be a point within $(a'b')$ such that $c' - a' = b - a$, $b' - c' = d - c$. For every pair of functions $x(t)$, $y(u)$ defined in (a, b) and (c, d) respectively, let us construct a new function $X(v)$ defined as follows in the interval (a', b'): for $v = a'$, or within the interval (a', c'), $X(v) = x(a + v - a')$; for $v = b'$, or within the interval (c', b'), $X(v) = y(c + v - c')$; for $v = c'$ we shall give $\overline{X}(c')$ one of the two values $x(b)$ or $y(c)$. Since

---

(^{1}) We can omit the letters a, b and simply write

$\mathrm{F}\,[x\,(t, \alpha, \beta \ldots)]$ and $\mathrm{F}\,[x\,(t);\, \lambda,\, \mu \ldots]$

when there is no ambiguity about the rôle of the variable t and of the other variables $\alpha, \beta, \ldots; \lambda, \mu, \ldots$.

the pair of functions $x(t)$, $y(u)$ completely determine the function $X(v)$ (with the sole exception of $v = c'$), and *vice versa* $\bar{X}(v)$ determines the two functions $x(t)$, $y(u)$ (with the exception of $t = b$ for x, and $u = c$ for y), we can say, so long as the indeterminateness of the variable function at an isolated point does not affect the value of the functional, that the quantity $z = \mathrm{F}\left[\underset{a}{x(t)}^{b}, \underset{c}{y(u)}^{d}\right]$ depends on all the values taken by the *single function* $\bar{X}(v)$ in the interval (a', b'), i. e. that

$$z = \mathrm{F}\left[\underset{a}{x(t)}^{b}, \underset{c}{y(u)}^{d}\right] = \mathrm{G}\left[\underset{a'}{\bar{X}(v)}^{b'}\right].$$

8. Another generalisation of the concept of a *functional* is obtained by considering those quantities z *which depend on all the values taken by one or more functions* $\varphi_1(x_1, x_2, \ldots x_n)$, $\varphi_2(y_1, y_2, \ldots y_m), \ldots$ *of several variables defined within certain fields* $C_1, C_2, \ldots$ *respectively*([1]):

$$z = \mathrm{F}[\varphi_1(x_1, x_2, \ldots x_n), \varphi_2(y_1, y_2, \ldots y_m), \ldots].$$

9. This concept of a functional also includes *functions of lines*, and, more generally, *of hyperspaces* ([2]). We say, in fact, that a quantity z is a function of a variable hyperspace S_r, contained in another S_n ($n > r$) and parametrically defined by the equations

$$x_i = \varphi_i(t_1, t_2, \ldots t_r), \quad i = 1, 2, \ldots n,$$

when to every such S_r there corresponds one definite value of the quantity $z = \mathrm{F}[S_r]$, which thus comes to depend on all the values taken by the n functions φ_i at the points of the field C (of r dimensions) for which they are defined. It is to be noted, however, that this function z, a function of the hyperspace S_r, is *not* a general functional of the n functions φ_i; in fact, if the parameters t_k are changed into others u_l by means of the transformation

$$t_k = t_k(u_1, u_2, \ldots u_r), \quad k = 1, 2, \ldots r,$$
$$\frac{d(t_1, t_2, \ldots t_r)}{d(u_1, u_2, \ldots u_r)} \neq 0,$$

([1]) Cf. FABRI, (19) and (20). (The numbers in parentheses in the footnotes refer to the bibliography at the end of each chapter.)
([2]) Cf. VOLTERRA, (83) and (84).

I GENERALITIES AND DEFINITIONS 7

the coordinates x_i will no longer be given by the functions φ of the t_k's but by other functions ψ of the u_l's:

$$x_i = \psi_i(u_1, u_2, \ldots u_r) = \varphi_i(t_1, t_2, \ldots t_r),$$

but the quantity z which depends *only on the form* of the hyperspace S_r (or manifold, as it is often called), and not on the mode of representation, will not have changed; i.e. z is a functional of the functions φ_i which does not vary when the φ_i's are replaced by other functions ψ_i obtained from them by a change of parameters.

§ 3. Functional Fields and Abstract Aggregates.

10. A functional $F[x(t)]$ of the function $x(t)$ will be defined in general only when $x(t)$ varies within a certain determinate field of functions. Thus, for instance, the functional

$$F[x(t)] = \int_a^b x(t)\,dt$$

is defined only for the field of the functions $x(t)$ that are *integrable* in the interval (a, b); the functional

$$F[x(t)] = \int_a^b f\left(t, x, \frac{dx}{dt}, \frac{d^2x}{dt^2}, \ldots \frac{d^n x}{dt^n}\right) dt$$

is defined only for those functions $x(t)$ for which derivatives exist up to the nth order, and for which

$$\varphi(t) = f(t, x(t), x'(t) \ldots x^{(n)}(t))$$

is integrable in the interval (a, b); and so on.

The study of *functional fields*, i.e. of those aggregates whose elements are *functions*, is therefore of the greatest interest for an accurate understanding of the concept of a functional. The widest possible functional field, that consisting of all possible functions defined in an interval (a, b), obviously constitutes a space of an infinite non-enumerable number of dimensions (i.e. the number of its dimensions is the same as the power of the continuum), since each of its elements $x(t)$ is defined only when the ∞ values taken by x corresponding to the ∞ values of t in the interval (a, b) are known. Very few functionals, however, are defined for the whole of this field. A much more restricted field, but of notable interest, is that of the *analytical functions* $\alpha(t)$ in the region round the real segment (a, b); this has an infinite, but *enumerable*, number of dimensions, since its elements $\alpha(t)$ are determined, e. g., by the values of α and its successive derivatives at a fixed point t_0. A wider field is that of the *continuous functions* $y(t)$ (also of an infinite but enumerable number of dimensions, since they are determined by the values at rational points only), whose elements $y(t)$ can always be considered as the *limits of analytical functions*, or even of polynomials (WEIERSTRASS) ([1]).

11. A still wider field is that composed of the functions that are *limits of continuous functions*; in general, we can adopt BAIRE's well-known classification([2]) of functions into categories, which, even if it excludes some of the more irregular functions, nevertheless includes all those that are of most interest for analysis.

Other interesting functional fields are those of functions *differentiable* to a certain order, functions of *limited variation*, functions *whose square is integrable*, etc.

12. A general theory, not only of functional fields, but of "abstract aggregates" whose elements may be any quantities whatever $A, B, \ldots$ not specified individually, has been given by MOORE and FRÉCHET (see below, Chapter VI, No. 164, and its bibliography) in a series of memoirs and in a treatise containing the basis of a new theory called *general analysis*.

Extending the concept of a *function*, or *functional operator*, to these abstract aggregates, we shall say that $U[A]$ is a function, or a functional, uniform in an aggregate Y, if to every element A of Y there can be made to correspond a completely determined number $U[A]$ ([3]).

If, in particular, given an aggregate Y we consider as elements of a new aggregate the partial aggregates $Y_1, Y_2, \ldots Y_h$ that can be extracted from Y, and if a number $V[Y_h]$ is made to correspond to each of these, we shall say that $V[Y_h]$ is a *function of an aggregate* (a generalisation of the concept of a *function of a line* (see above, No. 9), the line being a partial aggregate of points extracted from the space S_n in which it is immersed). Of particular importance among these are the

---

([1]) Cf. WEIERSTRASS, (90).
([2]) Cf. BAIRE, (7).
([3]) FRÉCHET extends this concept to the relations between two elements of any nature whatever.

I GENERALITIES AND DEFINITIONS 9

additive functions of aggregates $V[Y_h]$ which have the property that if Y_1, Y_2 are two aggregates with no common element, and if $\bar{Y} = Y_1 + Y_2$ is the aggregate formed by combining them, then the value of $V[\bar{Y}] = V[Y_1 + Y_2]$ is equal to the sum $V[Y_1] + V[Y_2]$. Such is, for instance, the *measure* $M[Y_h]$ of an aggregate Y_h of points of a segment which serves to define an integral in LEBESGUE's sense [1]. Another example is given by the increment $\Delta f = f(t_2) - f(t_1)$ of a function of limited variation in the interval (t_1, t_2) (a particular aggregate of points of a segment (a, b)); this function of an aggregate serves to define the STIELTJES integral [2],

$$\int_a^b x(t) \cdot df(t), \quad \text{as} \quad \lim_{n=\infty} \sum_r^n \bar{x}_r \Delta f_r,$$

when it exists, i.e. the limit of the sum of the products of the values of $\bar{x}_r$ (included between upper and lower limits of the function $x(t)$ in the interval (t_{r-1}, t_r)) by the increments $\Delta f_r = f(t_r) - f(t_{r-1})$ when the maximum amplitude μ of the n intervals (t_{r-1}, t_r), into which the total interval (a, b) is divided, tends to 0.

§ 4. *Continuity.*

14. Before we can proceed with the study of these abstract aggregates, however, we must lay down some hypotheses on the nature of their elements. Thus, for instance, to establish the fundamental concept of the *limiting element of a sequence*, FRÉCHET supposes that for any pair of elements A and B of the aggregate a number $(A, B) = (B, A) \geq 0$ (the distance between the two elements) can be defined such that: (i) (A, B) is not zero except when $A = B$; (ii) for three elements A, B, C, we have always $(A, B) \leq (A, C) + (C, B)$; the element A is then called the *limit* of the sequence $A_1, A_2, \ldots A_n \ldots$ (written as $A = \lim\limits_{n=\infty} A_n$) if the positive quantity (A, A_n) has zero for its limit when n tends to ∞. Similarly we can define the *limiting elements* of an aggregate as those elements which are the limits of a suitable sequence extracted from the aggregate.

---

[1] Cf. LEBESGUE, (55).
[2] Cf. P. LÉVY, (60).

If an aggregate contains the aggregate (first derivative) composed of all its limiting elements, it is said to be *closed*. If an aggregate contains a finite number of elements, or if it contains an infinite number and is such that any infinity of its elements always gives rise to at least one limiting element, the aggregate is said to be *compact* ([1]).

15. Having thus established the concept of a *limit*, we can now define the *continuity* of a functional $U[A]$. We shall say that the functional is *continuous at an element A (the limiting element of the aggregate Y within which* U *is defined)* if

whenever

$$\lim_{n=\infty} U[A_n] = U[A]$$

$$\lim_{n=\infty} A_n = A;$$

and we shall say that the functional is *uniformly continuous in a field* if it is not only continuous at each of its elements A, but also if given an arbitrary quantity $\varepsilon > 0$ we can determine a number $\eta > 0$ for which always

$$|U[A] - U[A']| < \varepsilon$$

for $(A, A') < \eta$, whatever A and A' may be. It must however be noted that a functional continuous at every element of an aggregate is not necessarily continuous in the aggregate itself, an important difference from what happens in the case of functions of a finite number of variables. The continuity will however be uniform if the aggregate is *compact* (No. 14) ([2]).

Lastly, it must be pointed out that since the continuity of a functional depends on the definition of the *limit* of a sequence contained in the aggregate Y, it depends ultimately on the more or less arbitrary definition that has been given of the *distance* (A, B) between two elements; so that any change in this definition may mean that a functional ceases to be continuous, or *vice versa*.

Suppose, for instance, that the aggregate Y in which the functional is defined is the aggregate of the functions $x(t)$, defined and differentiable in the interval (0, 1). We can take as the definition of the distance between two functions $x_1(t)$,

---

([1]) Cf. FRÉCHET, (28).
([2]) Ibid.

$x_2(t)$, the maximum value assumed by $|x_1(t) - x_2(t)|$ as t varies in the interval (0, 1). With this definition, the functional

$$F[x(t)] = \int_0^1 f(t, x(t)) dt,$$

where f is a continuous function of its arguments, is a *continuous functional* of $x(t)$, while the functional

$$G[x(t)] = \overline{\lim} \, x'(t) \, (^1)$$

is not a continuous functional of $x(t)$. In fact, if we put

$$x_1(t) = k, \quad x_2(t) = k + \varepsilon \sin \frac{2\pi t}{\varepsilon},$$

we shall have $|x_1(t) - x_2(t)| < \varepsilon$, and therefore, for our definition of distance,

$$\lim_{\varepsilon = 0} x_2(t) = x_1(t);$$

but we have also

$$x_1'(t) = 0, \quad x_2'(t) = 2\pi \cos \frac{2\pi t}{\varepsilon},$$

$$G[x_1(t)] = \overline{\lim} \, x_1'(t) = 0,$$

$$G[x_2(t)] = \overline{\lim} \, x_2'(t) = 2\pi \neq 0,$$

and therefore

$$\lim_{\varepsilon = 0} G[x_2(t)] = 2\pi$$

which is *distinct from* $G[x_1(t)]$. When a functional $U[x(t)]$ is continuous at the element $x(t)$ with this definition of the distance between two functions, it is said to have *continuity of order* 0 at the element $x(t)$. In many treatises (²) this continuity is called "uniform" continuity of order 0, with reference to the fact that if $\lim_{n=\infty} x_n(t) = x(t)$, the difference $|x_n(t) - x(t)|$ can be made as small as we please, *however the variable element t* (which is here a number) *is chosen* in its interval of variation (a, b). But in order to avoid confusion with the uniform continuity of functionals defined in this No. 15 (in which the

---

(¹) $\overline{\lim}$ is "the greatest of the limits".
(²) Cf. Lévy, (60).

variable element to be chosen arbitrarily is not a number, but is in general an element A which might e.g. be a function), we shall not use the adjective "uniform". In other words, in this case it is not the *functional* that has "uniform" continuity, but it is the variable element $x_n(t)$, considered as a function of t in the interval (a, b), that tends "uniformly" to $x(t)$. If on the contrary we adopted for the definition of the distance between two functions $x_1(t)$, $x_2(t)$ the upper limit of the values assumed both by $|x_1(t) - x_2(t)|$ and by $|x_1'(t) - x_2'(t)|$ in the interval considered, the resulting continuity would be of the *first order*, and so on. With this second definition, the functional $G[x(t)]$ previously defined, which *had not continuity of order* 0, *is now continuous (has continuity of order* 1*)*.

If we consider, lastly, the aggregate Y of the functions $x(t)$ defined in the interval $(0, 1)$ and whose squares are integrable, and if we agree not to consider as distinct two functions $x_1(t)$ and $x_2(t)$ whose difference $x_1(t) - x_2(t)$ differs from 0 only at an aggregate of points of measure zero, we can then adopt yet another definition of the distance ϱ between the two functions $x_1(t)$, $x_2(t)$, namely,

$$\varrho = \int_0^1 [x_1(t) - x_2(t)]^2 \, dt \; (^1).$$

In fact, if $\varrho = 0$, $x_1(t) - x_2(t) = 0$, except at most at an aggregate of points of measure zero. Continuous functionals with this last definition of distance are called *continuous on the average*.

16. Before concluding these considerations on the continuity of functionals, we may refer to a remark due to GATEAUX([2]) regarding an approximation to functionals with continuity of order 0 (No. 15 above). If we denote by $F[y(t)]$ such a functional defined in the field of real and continuous functions $y(t)$ ($0 \leq t \leq 1$, $A \leq y \leq B$), it will be an ordinary function of the n variables $y_1, y_2, \ldots y_n$ when $y(t)$ is replaced by a function $\eta(t)$ defined as follows:

$$\eta(t) = y_1 = y\left(\frac{1}{n}\right) \text{ for } 0 \leq t \leq \frac{1}{n}, \quad \eta\left(\frac{h}{n}\right) = y_h = y\left(\frac{h}{n}\right),$$

([1]) Cf. LÉVY, (60).
([2]) Cf. GATEAUX, (43). The unpublished papers of this young French mathematician who died in the great war were edited by P. LÉVY in the years 1919 and 1922. Cf. (43), (44), (45).

I GENERALITIES AND DEFINITIONS 13

with $\eta(t)$ varying linearly between $y\left(\dfrac{h}{n}\right)$ and $y\left(\dfrac{h+1}{n}\right)$ in the interval $\dfrac{h}{n} \leq t \leq \dfrac{h+1}{n}$ $(h = 1, 2, 3, \ldots, n-1)$.

Putting
$$F[\eta(t)] = f_n(y_1, y_2, \ldots y_n)$$

and remembering that $\lim_{n=\infty} \eta(t) = y(t)$ since $y(t)$ is continuous, and that the functional is continuous of order 0, we can easily show that

$$F[y(t)] = \lim_{n=\infty} f_n(y_1, y_2, \ldots y_n)$$

and the convergence is uniform in every *compact* aggregate of *continuous* functions.

Continuous *functionals* of order 0 are thus expressed as *limits of functions* of n variables, when the number n of the variables increases towards ∞.

17. Extremes of a functional. Many of the theorems that hold for functions of real variables can be extended to functionals defined in an abstract aggregate. Of special importance for the applications it has in the calculus of variations is a theorem analogous to that of WEIERSTRASS. This theorem, which has been proved in its most general form by FRÉCHET [1], is as follows:

Every functional U [A] *which is uniform and continuous in a compact closed aggregate* Y (cf. No. 14): (i) *is limited in* Y; (ii) *has the value of its upper limit (maximum) at least once in* Y.

For continuous functionals of order 0 defined within *closed* aggregates of *functions* $x(t)$ which are themselves also continuous (i.e. such that the ratio of the increments $\dfrac{x(t_1) - x(t_2)}{t_1 - t_2}$ remains always between two finite numbers l, L), a fundamental theorem of this kind was proved as far back as 1895 by ARZELÀ [2], together with many other analogous theorems.

More recently HADAMARD, in his treatise on the calculus of variations [3], has also treated this calculus as a particular

---

[1] Cf. FRÉCHET. (27).
[2] Cf. ARZELÀ, (2).
[3] Cf. HADAMARD, (48).

chapter in the more general "functional calculus", applying to it the general methods of the latter.

18. For the functionals which concern the calculus of variations, however, continuity of order 0 is too restrictive a condition and would lead to the exclusion of many of them. TONELLI([1]) has for this reason extended to functionals the definition of *lower* (or upper) *semicontinuity* already given by BAIRE for ordinary functions. According to this definition, we say that the functional F [A] is *semicontinuous from below* (or from above) at the element A if, given an arbitrary quantity $\varepsilon > 0$, we can determine a quantity $\varrho > 0$ such that

$$F[A'] > F[A] - \varepsilon \quad (\text{or } F[A'] < F[A] + \varepsilon)$$

for every A' for which $(A, A') < \varrho$.

A functional that is semicontinuous either from below or from above is a continuous functional in the ordinary sense (with a given definition of the distance (A, A')). Semicontinuous functionals form a large part of those ordinarily considered in the calculus of variations, and TONELLI has extended to them the foregoing existence theorems for maxima and minima, thus definitely setting the calculus of variations in its natural place in the functional calculus, of which it forms one of the most important and most ancient chapters, and making it independent of the theory of differential equations.

§ 5. *Linear Functionals.*

19. After these general considerations, we shall proceed to consider some cases of particularly interesting functionals, and, in the first place, the so-called *linear functionals* or functionals of the *first degree*. These functionals are derived from linear forms of n variables, $P_1(y_i) = \sum_{1}^{n} k_i y_i$ by applying the usual method explained above (No. 3) for passing from a finite to an infinite number of variables; they will therefore be of the type

$$F_1\begin{bmatrix} b \\ y(t) \\ a \end{bmatrix} = \int_a^b k(t) y(t)\, dt, \tag{1}$$

where $k(t)$ is a given function and $y(t)$ the variable function.

([1]) Cf. TONELLI, (76) and (77).

If we put
$$y(t) = \lambda y_1(t) + \mu y_2(t) \tag{2}$$
we shall have
$$F_1[y(t)] = \lambda F_1[y_1(t)] + \mu F_1[y_2(t)]; \tag{3}$$
hence the name of *linear functionals* or *functionals of the first degree*. These functionals are continuous on the average ([1]). They are not however the only ones that have the property expressed by (3); thus, for example, the functional

$$U[y(t)] = \int_a^b k(t)\,y(t)\,dt + \Sigma_i \alpha_i y(\tau_i)$$

(where τ_i represents values between a and b) satisfies the condition (3), and is therefore a linear functional with continuity of order 0 if $\Sigma_i |\alpha_i|$ is convergent, but it cannot be put in the form (1) of an integral. The part represented by the ordinary integral is called the *regular part*, that represented by the summation term the *exceptional part*, and we also say that in this case the functional depends *specially* on the values taken by y at the *exceptional* points τ_i; the finite contribution $\alpha_i y(\tau_i)$ made by each of these to the functional is in fact of higher order of magnitude than the infinitesimal $k(t)\,y(t)\,dt$ contributed by another value taken at a point that is not exceptional.

20. A general expression for linear functionals of any order of continuity has been given by HADAMARD ([2]) in 1903. Since they are linear, we have in fact that if $y(t) = \Sigma_i k_i y_i(t)$, then

$$U[y(t)] = \Sigma_i k_i U[y_i(t)],$$

and also, passing from a sum to an integral,

$$U\left[\int_a^b f(\lambda)\,y_\lambda(t)\,d\lambda\right] = \int_a^b f(\lambda)\,U[y_\lambda(t)]\,d\lambda.$$

In particular, if we put

$$y_\mu(t) = \frac{\int_0^1 e^{-\mu^2(\lambda-t)^2} y(\lambda)\,d\lambda}{\int_0^1 e^{-\mu^2(\theta-t)^2}\,d\theta}$$

---

[1] Cf. FRÉCHET, (26), and STEINHAUS, (7.).
[2] Cf. HADAMARD, (47) and (48).

(in this case the former $f(\lambda)$ is now $y(\lambda)$), we shall get

$$U[y_\mu(t)] = \int_a^b y(\lambda) F(\lambda,\mu) d\lambda,$$

where we have put

$$F(\lambda,\mu) = U\left[\frac{e^{-\mu^2(\lambda-t)^2}}{\int_0^1 e^{-\mu^2(\theta-t)^2} d\theta}\right]$$

(which is independent of the variable function y). But since $y_\mu(t)$ tends uniformly to $y(t)$ as μ increases towards ∞, we shall have, from the assumed continuity of the functional,

$$U[y(t)] = \lim_{\mu=\infty} U[y_\mu(t)] = \lim_{\mu=\infty} \int_a^b F(\lambda,\mu) y(\lambda) d\lambda.$$

This is HADAMARD's general formula; similar expressions could also be found for linear functionals (limits of integrals) by taking the function $F(\lambda,\mu)$ in an infinite number of other forms.

21. Another expression for all possible functionals $U[y(t)]$ which are linear (i.e. satisfy condition (3)), continuous of order 0, and limited in the field of functions that are continuous in (a, b) or have a finite number of discontinuities of the first kind, is obtained as follows.

Let $y_\tau(t)$ be the function that $= 1$ for $a \leq t \leq \tau$, and 0 for $\tau < t \leq b$, and put

$$U[y_\tau(t)] = f(\tau) \tag{4}$$

(cf. No. 6); it can easily be seen that the function $f(\tau)$ is *of limited variation*, since we have supposed the functional to be limited. Now let us replace any function whatever $y(t)$ of the field by another function $y_n(t)$ defined as follows: divide the interval (a, b) into n partial intervals (t_{i-1}, t_i), and in each of these place $y_n(t)$ equal to one of the values $y_n(\bar{t}_i)$ taken by y in the same partial interval; then $y_n(t)$ will tend uniformly to $y(t)$ when the maximum amplitude of the partial intervals tends to 0. But since the functional is linear, we have

$$U[y_n(t)] = \sum_{i=1}^n y_n(\bar{t}_i)[f(t_i) - f(t_{i-1})],$$

I GENERALITIES AND DEFINITIONS 17

and as we have supposed that U is continuous,

$$U[y(t)] = \lim U[y_n(t)] = \lim \sum_i^n y_n(t_i)[f(t_i) - f(t_{i-1})].$$

It follows that the limit of the sum exists when the maximum amplitude of the partial intervals tends to 0, and this limit is precisely equal to $U[y(t)]$. Further, this limit is also, by definition, the STIELTJES integral

$$\int_a^b y(t) df(t) = U[y(t)],$$

and this is consequently the most general expression ([1]) for linear functionals that are continuous of order 0, and limited; which are thus determined by the *characteristic function* $f(t)$ given by (4).

If $f(t)$, defined by (4), is continuous and differentiable in the whole interval (a, b), the functional reduces to a regular functional

$$U[y(t)] = \int_a^b y(t) df(t) = \int_a^b f'(t) y(t) dt.$$

If however $f(t)$ has discontinuities (necessarily of the first species, since it is of limited variation) at a finite number (or an enumerable infinity) of points, and if we denote by α_i the leap of the function at these points of discontinuity, the function remaining differentiable at all other points, the functional takes the form

$$U[y(t)] = \int_a^b f'(t) y(t) dt + \sum_i \alpha_i y(\tau_i),$$

i.e. it depends *specially* on the values of y at the points of discontinuity of $f(t)$. Lastly, if $f(t)$ has no derivatives at a set of points which is non-enumerable but of measure zero (as well as at the points of discontinuity) — the general case — then the linear functional can be decomposed into the sum of three terms

$$U[y(t)] = \int_a^b f'(t) y(t) dt + \sum_i \alpha_i y(\tau_i) + \int_a^b y(t) df_3(t),$$

---

[1] Cf. RIESZ, (72).

of which the first is a regular functional, the second is the exceptional part, and the third is the really new element introduced by the STIELTJES integral (determined by the continuous function $f_3(t)$ which has its derivative zero everywhere except at the non-enumerable aggregate of measure zero).

22. Other linear functions, in a certain sense more general, with continuity of order p (all that are continuous of order 0 are also continuous of order p, but not *vice versa*), are those represented by

$$U[y(t)] = \int_a^b k(t) y(t) dt + \sum_i \alpha_i y(\tau_i) + \sum_i \beta_i y'(\tau_i) + \cdots$$
$$+ \sum_i \gamma_i y^{(p)}(\tau_i),$$

defined for functions that have derivatives to the pth order; we shall say that they *depend specially* on the values of y and of its derivatives to the pth order at the points τ_i. Since the functions $y(t)$ for which these functionals are defined can be put in the form

$$y(t) = y(\tau) + (t-\tau) y'(\tau) + \cdots + \frac{(t-\tau)^{p-1}}{(p-1)!} y^{(p-1)}(\tau)$$
$$+ \int_\tau^t \frac{(t-s)^{p-1}}{(p-1)!} y^{(p)}(s) ds$$

(τ being arbitrary in the interval (a, b) in which $f(t)$ is defined), we shall have, as the general formula representing all linear functionals that are limited and continuous of order p,

$$U[y(t)] = a_0 y(\tau) + a_1 y'(\tau) + \cdots$$
$$+ a_{p-1} y^{(p-1)}(\tau) + \int_a^b y^{(p)}(t) df(t),$$

where

$$a_i = U\left[\frac{(t-\tau)^i}{i!}\right] \quad \text{and} \quad f(t) = U[y_t(s)],$$

with

$$y_t(s) = \frac{(s-\tau)^p}{p!} \quad \text{for} \quad a \leq s \leq t,$$

$$y_t(s) = 0 \quad \text{for} \quad t < s \leq b \,(^1).$$

(^{1}) Cf. LÉVY, (60).

§ 6. *Functionals of Higher Degree.*

23. At the beginning of the last section we passed from the consideration of linear forms in n variables to that of linear functionals; analogously we can pass from the homogeneous functions of the second degree in n variables

$$P_2 = \sum_{rs} k_{rs} y_r y_s \qquad (1)$$

to the functionals $F_2[y(t)]$, which we shall call *regular homogeneous functionals of the second degree*, given by the general expression

$$F_2[y(t)] = \int_a^b \int_a^b k(\xi, \eta) y(\xi) y(\eta) d\xi d\eta; \qquad (2)$$

and just as in (1) we can always suppose $k_{rs} = k_{sr}$ (for if not we need only put $P_2 = \sum_{rs} k'_{rs} y_r y_s$, with $k'_{rs} = \dfrac{k_{rs} + k_{sr}}{2}$), so in (2) we can always suppose that $k(\xi, \eta) = k(\eta, \xi)$ (i.e. that the function $k(\xi, \eta)$, which determines the functional, is symmetrical in the two variables ξ, η); for if not we could replace the expression (2) by the equivalent

$$F_2[y(t)] = \int_a^b \int_a^b k'(\xi, \eta) y(\xi) y(\eta) d\xi d\eta$$

in which $k'(\xi, \eta) = \dfrac{k(\xi, \eta) + k(\eta, \xi)}{2}$, and therefore $k'(\xi, \eta) = k'(\eta, \xi)$.

More generally, we shall apply the term *regular homogeneous functionals of degree n* (an extension of *forms* of degree n to the case of an infinite number of variables) to the functionals given by the general expression

$$F_n[y(t)] = \int_a^b \int_a^b \cdots \int_a^b k(\xi_1, \xi_2, \ldots \xi_n) y(\xi_1) y(\xi_2) \cdots$$
$$\cdots y(\xi_n) d\xi_1 d\xi_2 \ldots d\xi_n, \qquad (3)$$

in which, by similar considerations to those used in the preceding case, we can always suppose that the function $k(\xi_1, \xi_2, \ldots \xi_n)$, the characteristic or kernel of the functional, is symmetrical with respect to its n variables.

We shall then apply the term *regular functionals of degree n* to the functionals

$$G_n[y(t)] = k_0 + F_1[y(t)] + F_2[y(t)] + \cdots + F_n[y(t)] \quad (4)$$

which are sums of regular homogeneous functionals whose highest degree is equal to n (an extension of *polynomials* of degree n to the case of an infinite number of variables).

24. A notable property of these regular functionals of degree n, $G_n[y(t)]$, is that they can be used to obtain an approximation to any continuous functional $G[y(t)]$ whatever. More precisely, we have the following theorem[1]:

Every functional $G[y(t)]$ continuous in the field of continuous functions can be represented by the expression

$$G\left[\underset{a}{\overset{b}{y(t)}}\right] = \lim_{n=\infty} G_{r_n}\left[y(t)\right]$$

i.e.

$$G\left[\underset{a}{\overset{b}{y(t)}}\right] = \lim_{n=\infty}\left[k_{n,0} + \int_a^b k_{n,1}(\xi)y(\xi)d\xi \right.$$
$$+ \int_a^b\int_a^b k_{n,2}(\xi_1,\xi_2)y(\xi_1)y(\xi_2)\,d\xi_1 d\xi_2 + \ldots$$
$$+ \int_a^b\int_a^b\cdots\int_a^b k_{n,r_n}(\xi_1,\xi_2,\ldots\xi_{r_n})y(\xi_1)y(\xi_2)\cdots$$
$$\left.\ldots y(\xi_{r_n})\,d\xi_1 d\xi_2\ldots d\xi_{r_n}\right], \quad (5)$$

where the functions $k_{n,r_n}(\xi_1,\xi_2,\ldots\xi_{r_n})$ are continuous functions, determined for the functional G, independently of the variable function $y(t)$.

This theorem generalises WEIERSTRASS's theorem[2] on continuous functions (which can be represented as limits of polynomials) to the case of an infinite number of variables. Further, the expansion given in (5) is uniformly convergent in every compact aggregate (cf. No. 14) of continuous functions (*compact* aggregates take the place in functional fields of *finite intervals* in one-dimensional fields); this corresponds to the

---

[1] Cf. FRÉCHET, (30).
[2] Cf. WEIERSTRASS, (90).

result in WEIERSTRASS's theorem that the polynomials converge *uniformly* to their limit if the functions are defined in a finite interval.

§ 7. *Functional Power Series.*

25. Instead of sums of homogeneous functionals finite in number, we can consider a series

$$G\left[y(t)\atop a\right]^b = \sum_{n}^{\infty} F_n[y(t)] = k_0 + \int_a^b k_1(\xi)y(\xi)d\xi$$

$$+ \int_a^b \int_a^b k_2(\xi_1, \xi_2) y(\xi_1) y(\xi_2) d\xi_1 d\xi_2 + \cdots$$

$$+ \int_a^b \int_a^b \cdots \int_a^b k_n(\xi_1, \xi_2, \ldots \xi_n) y(\xi_1) y(\xi_2) \ldots$$

$$\ldots y(\xi_n) d\xi_1 d\xi_2 \ldots d\xi_n + \cdots, \qquad (1)$$

convergent if $|y(t)| < \varrho$ (the radius of safe convergence), which makes a definite value of the functional $G\left[y(t)\atop a\right]^b$ correspond to every function $y(t)$ for which it converges. The functional so determined will not be of finite degree, but can be considered as obtained from power series in several variables (and therefore from analytical functions of several variables) when the number of these variables increases indefinitely.

These functionals, which can be represented by the series (1), which we shall also call *functional power series*, have for this reason been called *analytical functionals* by many writers ([1]). They are continuous of order 0; hence if the variable function y, within the region of convergence, depends analytically on a parameter α, i.e. if $y = y(\xi, \alpha)$ (an analytica function of α), then the functional $G[y(\xi, \alpha)] = f(\alpha)$, being the sum of a uniformly convergent series of analytical functions of α, is itself also an *analytical function of* α. Thus we can say that these functionals *maintain their analytical character* with respect to the parameter α. This fundamental property will be adopted further on (Chapter VI, No. 161) as the characteristic property for the definition of analytical functionals.

([1]) Cf. VOLTERRA, (83), and FRÉCHET, (30).

Section II. Operations on Functionals.

§ 1. *Differentiation and Derivation.*

26. Given the above definitions and examples, the problem now presents itself of establishing a general *calculus of functionals*, analogous to the ordinary calculus used for operations on functions. This will be possible by means of the introduction of suitable operations to be applied to functionals, and therefore belonging to a higher category than the functionals themselves.

Given a functional $F\left[\underset{a}{\overset{b}{y(t)}}\right]$, we shall begin by defining for it the operation of *differentiation*. For the functions of n variables $f(y_1, y_2, \ldots y_n)$ the total differential df is defined as

$$df = \sum_{1}^{n}{}_i \frac{\partial f}{\partial y_i} dy_i.$$

This quantity df has two fundamental properties:
(i) it is a linear form of the n quantities dy_i;
(ii) if we put $dy_i = \varepsilon\eta_i$, and take ε as principal infinitesimal, df differs by infinitesimals of higher order than the first from the increment of f corresponding to the increments $\varepsilon\eta_i$ of the variables.

Passing from the case of functions of n variables to the case of the functionals $F\left[\underset{a}{\overset{b}{y(t)}}\right]$, let us give $y(t)$ an increment $\delta y(t) = \varepsilon\eta(t)$, and let us try to define a quantity δF, corresponding to this increment, which shall have the two following properties:

(i) it is a linear functional of $\delta y(t) = \varepsilon\eta(t)$;

(ii) it differs by infinitesimals of higher order than ε from the increment ΔF of the functional corresponding to the increment $\delta y(t)$ of the variable function.

27. It was proved as early as 1887 ([1]) that this is certainly possible when the functional is *derivable* (has a first derivative) at every point in the interval (a, b). To define the first derivative at a point ξ of a functional F continuous of order 0, we shall follow a method analogous to that used for the functions

---

([1]) Cf. VOLTERRA, (83) and (84).

$f(y_1, y_2, \ldots y_n)$ of n variables. For these functions $\dfrac{\partial f}{\partial y_i}$ is defined as the limit, when it exists, of the ratio $\dfrac{\Delta_i f}{h}$, where h is an increment of the *variable* y_i *alone*, and $\Delta_i f$ is the increment of f corresponding to this increment h. For the functionals we shall give $y(t)$ an increment $\delta y(t) = \vartheta(t)$ which does not change sign, and such that $|\vartheta(t)| < \varepsilon$, $\vartheta(t) = 0$ outside an interval (m, n) of (a, b) of amplitude h, containing ξ in its interior; and we shall suppose:

(i) that the ratio $\dfrac{\Delta F}{\varepsilon h}$ is always less than a finite number M;

(ii) that, putting $\sigma = \int_m^n \vartheta(t)\,dt$, there exists a determinate and finite limit of $\dfrac{\Delta F}{\sigma}$ when ε and h tend simultaneously to 0, subject to the condition that the interval (m, n) always contains the point ξ in its interior;

(iii) that the ratio $\dfrac{\Delta F}{\sigma}$ tends to its limit uniformly with respect to all possible functions $y(t)$ and to all points ξ.

The limit of this ratio will then in general depend on the function $y(t)$; it will therefore be a *functional* of $y(t)$, and will also depend on the parameter ξ and will be a function of ξ. It will be denoted by the symbol

$$F'[y(t); \xi]$$

and will be called the *first derivative of the functional* F *with respect to the function* $y(t)$ *at the point* ξ. Just as for the functions $f(y_1, y_2, \ldots y_n)$ of n variables we had n partial derivatives $\dfrac{\partial f}{\partial y_i}$, depending on the *discontinuous index* i as well as on the n variables, so for the functionals we have an infinite number of derivatives depending both on the variable $y(t)$ and also on all the possible values within (a, b) of the *continuous parameter* ξ. Further, for derivable functions the total differential df is given by the formula

$$df = \sum_i \dfrac{\partial f}{\partial y_i} dy_i; \tag{1}$$

and it can be shown ([1]) analogously that for functionals $F[y(t)]$ that have a derivative $F'[y(t); \xi]$ continuous with respect to ξ and continuous of order 0 with respect to $y(t)$, it is possible to construct a quantity

$$\delta F = \int_a^b F'[y(t); \xi]\delta y(\xi)d\xi \qquad (2)$$

which satisfies precisely the two conditions laid down above. We shall therefore call it the *differential* or *first variation* of the functional F.

28. The result just found can also be expressed in another form. In fact putting $\delta y(t) = \varepsilon \varphi(t)$, we can say, with the foregoing hypotheses, that

$$\lim_{\varepsilon=0} \frac{\Delta F}{\varepsilon} = \left(\frac{d}{d\varepsilon} F[y(t) + \varepsilon\varphi(t)]\right)_{\varepsilon=0} = \int_a^b F'[y(t); \xi]\varphi(\xi)d\xi \,([2]). \quad (3)$$

§ 2. *Derivatives of the Second and Higher Orders.*

29. If the functional $F'[y(t); \xi_1]$ is in its turn derivable, we shall denote its first derivative at the point ξ_2 by $F''[y(t); \xi_1, \xi_2]$, which will be called the *second derivative* of the functional F at the points ξ_1 and ξ_2. When $F''[y(t); \xi_1, \xi_2]$ is a continuous functional of order 0 with respect to $y(t)$, and a continuous function with respect to ξ_1 and ξ_2, it can be shown ([1]) that

$$F''[y(t); \xi_1, \xi_2] = F''[y(t); \xi_2, \xi_1],$$

so that it is a symmetric function of the two parameters ξ_1 and ξ_2. This property is a generalisation of the well-known property of functions $\dfrac{\partial^2 f}{\partial y_i \partial y_k} = \dfrac{\partial^2 f}{\partial y_k \partial y_i}$ (the mixed derivative depends symmetrically on the indices; theorem on the inversion of the order of differentiation); it can also be expressed by saying that the derivative at the point ξ_2 of the derivative at the point ξ_1 is equal to the derivative at the point ξ_1 of the derivative at the point ξ_2, or by saying that the result is unchanged if the order of the two differentiations is interchanged.

([1]) Cf. VOLTERRA, (83).
([2]) Cf. the recent paper by R. COURANT, (9).

I OPERATIONS ON FUNCTIONALS 25

Applying the last formula of the preceding section to F' we shall have

$$\left(\frac{d}{d\varepsilon} F'[y(t) + \varepsilon \varphi(t); \xi_1]\right)_{\varepsilon=0} = \int_a^b F''[y(t); \xi_1, \xi_2] \varphi(\xi_2) d\xi_2;$$

$$\left(\frac{d^2}{d\varepsilon^2} F[y(t) + \varepsilon \varphi(t)]\right)_{\varepsilon=0}$$
$$= \left(\frac{d}{d\varepsilon} \int_a^b F'[y(t) + \varepsilon\varphi(t); \xi_1] \varphi(\xi_1) d\xi_1\right)_{\varepsilon=0};$$

and therefore

$$\left(\frac{d^2}{d\varepsilon^2} F[y(t) + \varepsilon \varphi(t)]\right)_{\varepsilon=0}$$
$$= \int_a^b \int_a^b F''[y(t); \xi_1, \xi_2] \varphi(\xi_1) \varphi(\xi_2) d\xi_1 d\xi_2.$$

For the derivatives of higher order $F^{(n)}[y(t); \xi_1, \xi_2, \ldots \xi_n]$ it can be seen similarly that with the usual hypotheses as to continuity they are *symmetric functions of the n parameters* ξ_1, $\xi_2, \ldots \xi_n$, and also that

$$\left(\frac{d^n}{d\varepsilon^n} F_t[y(t) + \varepsilon \varphi(t)]\right)_{\varepsilon=0} = \int_a^b \int_a^b \ldots$$
$$\ldots \int_a^b F^{(n)}[y(t); \xi_1, \xi_2, \ldots \xi_n] \varphi(\xi_1) \varphi(\xi_2) \ldots \varphi(\xi_n) d\xi_1 d\xi_2 \ldots d\xi_n.$$

These nth derivatives with respect to the parameter ε are therefore expressed by means of regular functionals of the nth degree of the function $\varphi(\xi)$ when $y(t)$ is considered fixed.

§ 3. *Extension of Taylor's Theorem*[1].

30. If in the functional $F[y(t) + \varepsilon\varphi(t)]$ we consider $y(t)$ and $\varphi(t)$ fixed, the functional itself becomes an ordinary function of ε,

$$f(\varepsilon) = F[y(t) + \varepsilon \varphi(t)],$$

and if the functional is derivable up to the nth order then $f(\varepsilon)$,

---

[1] Cf. VOLTERRA, (83).

by the results reached above, will also be differentiable up to the nth order. Applying Taylor's theorem to it, we shall then have

$$f(1) = f(0) + \left(\frac{\partial f}{\partial \varepsilon}\right)_{\varepsilon=0} + \cdots + \frac{1}{(n-1)!}\left(\frac{\partial^{n-1} f}{\partial \varepsilon^{n-1}}\right)_{\varepsilon=0}$$
$$+ \frac{1}{n!}\left(\frac{\partial^n f}{\partial \varepsilon^n}\right)_{\varepsilon=\theta},$$

where θ lies between 0 and 1; and, by the preceding formulae,

$$F[y(t) + \varphi(t)] = F[y(t)] + \sum_{1}^{n-1} \frac{1}{i!} \int_a^b \int_a^b \cdots$$
$$\cdots \int_a^b F^{(i)}[y(t); \xi_1, \xi_2, \ldots \xi_i] \varphi(\xi_1) \varphi(\xi_2) \cdots \varphi(\xi_i) d\xi_1 d\xi_2 \ldots d\xi_i$$
$$+ \cdots + \frac{1}{n!} \int_a^b \int_a^b \cdots \int_a^b F^{(n)}[y(t)$$
$$+ \theta \varphi(t); \xi_1, \xi_2, \ldots \xi_n] \varphi(\xi_1) \varphi(\xi_2) \cdots \varphi(\xi_n) d\xi_1 d\xi_2 \ldots d\xi_n,$$

a formula which extends the ordinary form of Taylor's theorem for functions of n variables to derivable functionals. If the functional has derivatives of any order whatever, and if the limit of the last term is 0 for $n = \infty$, we shall have

$$F[y(t) + \varphi(t)] = F[y(t)] + \sum_{1}^{\infty} \frac{1}{n!} \int_a^b \int_a^b \cdots$$
$$\cdots \int_a^b F^{(n)}[y(t); \xi_1, \xi_2, \ldots \xi_n] \varphi(\xi_1) \varphi(\xi_2) \cdots \varphi(\xi_n) d\xi_1 d\xi_2 \ldots d\xi_n.$$

If we suppose $y(t)$ fixed, this formula gives us an expression in terms of definite integrals of the functional

$$G[\varphi(t)] = F[y(t) + \varphi(t)],$$

which is therefore analytical (No. 25).

§ 4. *Other Differentiable Functionals.*

31. The derivable functionals considered above do not, however, exhaust the class of functionals $F[y(t)]$ that admit of a differential. In the preceding cases, the differential δF was

a *regular* linear functional of the increment $\delta y(t)$; but there can also exist functionals, e.g. in the calculus of variations, with a variation δF that is a linear functional of $\delta y(t)$, but depending *specially* (cf. No. 19) on the values of δy and of its derivatives at some points of the interval (at these points the derivative of the functional, in the above sense, does not exist, or, better, becomes infinite).

Denoting by $\delta F[y(t), \varphi(t)]$ the first variation or differential of the functional $F[y(t)]$ corresponding to the increment $\delta y(t) = \varphi(t)$ of y, δF is a functional of the two functions $y(t)$ and $\varphi(t)$), and remembering the two fundamental properties of variations (No. 26), we shall have in general, *for every differentiable functional*,

$$F[y(t) + \varphi(t)] = F[y(t)] + \delta F[y(t), \varphi(t)] + \eta L(\varphi), \quad (1)$$

where $L(\varphi)$ denotes the distance of $\varphi(t)$ from the particular function $\overline{\varphi} = 0$, and η a quantity which tends to 0 with $L(\varphi)$. It follows from this that if the variation is a linear functional of φ *continuous of order* μ, then the functional $F[y(t)]$ will also be continuous of order μ ([1]).

If in (1) we put $\varepsilon \varphi(t)$ instead of $\varphi(t)$ and take the derivative with respect to ε, taking into account that δF is linear with respect to the second variable function, we shall get

$$\left(\frac{d}{d\varepsilon} F[y(t) + \varepsilon \varphi(t)]\right)_{\varepsilon=0} = \delta F[y(t), \varphi(t)], \quad (2)$$

which in the case of δF a *regular* linear functional of $\varphi(t)$ reduces to the formula (3) of No. 28 already found.

32. More generally, it can be shown ([2]) if $y(t, \alpha)$ is a function differentiable with respect to α, and if also $y(t, \alpha)$ and $y'_\alpha(t, \alpha) = \dfrac{\partial y}{\partial \alpha}$ are continuous in the field $\alpha_1 \leq \alpha \leq \alpha_2$, $a \leq t \leq b$, while the functional $F\left[\overset{b}{\underset{a}{y(t)}}\right]$ is differentiable, then the derivative $\dfrac{df}{d\alpha}$ of the function $f(\alpha) = F[y(t, \alpha)]$ exists for α included in

---

([1]) Cf. FRÉCHET, (31).
([2]) Ibid.

the interval (α_1, α_2), and we have precisely

$$\frac{df}{d\alpha} = \frac{d}{d\alpha} F[y\overset{b}{\underset{a}{(t,\alpha)}}] = \delta F[y(t,\alpha), y'_\alpha(t,\alpha)], \qquad (3)$$

a formula which generalises for functionals the rule for differentiation of the compound functions $f(y_1, y_2, \ldots y_n)$ with $y_i = y_i(\alpha)$, namely,

$$\frac{df}{d\alpha} = \sum_i \frac{\partial f}{\partial y_i} \frac{dy_i}{d\alpha}. \qquad (4)$$

In particular, in the case of derivable functions $F[y(t)]$ (δF a *regular* linear functional of the second variable function), (3) becomes

$$\frac{df}{d\alpha} = \frac{d}{d\alpha} F[y(t,\alpha)] = \int_a^b F'[y(t,\alpha); \xi] y'_\alpha(\xi, \alpha) d\xi,$$

which we can suppose obtained from (4) by the usual method (No. 3) of passing from a finite to an infinite number of variables.

33. From (3) it also follows that if the functional $F[y(t)]$ has a relative maximum (or minimum) at $y_0(t) = y(t, \alpha_0)$, i.e. if always $F[y_0(t)] > F[y(t)]$ for $y(t)$ sufficiently near to $y_0(t)$, then the quantity

$$\frac{df}{d\alpha} = \frac{d}{d\alpha} F[y(t,\alpha)] = \delta F[y(t,\alpha), y'_\alpha(t,\alpha)]$$

must vanish for $\alpha = \alpha_0$, $y'_\alpha(t, \alpha_0)$ being an arbitrary function of t; or, in other words, *the first differential* $\delta F[y(t), \varphi(t)]$ *of a functional is zero for those functions* $y_0(t)$ *which make the functional itself a maximum or minimum* ([1]). For derivable functionals we shall have in consequence that *the derived functional* $F'[y(t); \xi]$ *is identically zero, whatever ξ may be, for those functions $y_0(t)$ for which the functional is a maximum or minimum*.

§ 5. *Calculation of Some Variations.*

34. If we take a *regular* functional *of degree n* (No. 23)

$$F[y(t)] = k_0 + \sum_1^n \int_a^b \int_a^b \ldots \int_a^b k_i(t_1, t_2, \ldots t_i) y(t_1) y(t_2) \ldots$$
$$\ldots y(t_i) dt_1 dt_2 \ldots dt_i,$$

([1]) Cf. VOLTERRA, (83).

its variation δF, corresponding to the increment $\delta y(t)$ of the variable function, will be

$$\delta F = \sum_{1}^{n} i \int_a^b \delta y(\xi)\, d\xi \int_a^b \int_a^b \ldots$$
$$\ldots \int_a^b k_i(t_1, t_2, \ldots t_{i-1}, \xi) y(t_1) y(t_2) \ldots y(t_{i-1})\, dt_1\, dt_2 \ldots dt_{i-1},$$

and therefore its first derivative at the point ξ, namely,

$$F'[y(t); \xi] = k_1(\xi) + \sum_{2}^{n} i \int_a^b \int_a^b \ldots$$
$$\ldots \int_a^b k_i(t_1, t_2, \ldots t_{i-1}, \xi) y(t_1) y(t_2) \ldots y(t_{i-1})\, dt_1\, dt_2 \ldots dt_{i-1},$$

will be a *regular functional of degree* $n-1$. In general, its pth derivative, for $0 < p < n$, will be a regular functional of degree $n-p$, while for $p = n$ it reduces to

$$F^{(n)}[y(t); \xi_1, \xi_2, \ldots \xi_n] = n!\, k_n(\xi_1, \xi_2, \ldots \xi_n),$$

a function of the parameters ξ alone and independent of $y(t)$; for $p > n$ it will always be $F^{(p)}[y(t); \xi_1, \xi_2, \ldots \xi_p] = 0$, analogously to what happens in the case of polynomials of degree n. *Vice versa*, if at all points of the interval (a, b) a functional has a derivative of order p which is a regular functional of degree $n-p$, then it will itself be a regular functional of degree n.

35. If instead we consider the non-regular functional ([1]) of the second degree

$$F[y(t)] = k_0 + \int_a^b k_1(t) y(t)\, dt + a \int_a^b y^2(t)\, dt$$
$$+ \int_a^b \int_a^b k_2(t_1, t_2) y(t_1) y(t_2)\, dt_1\, dt_2,$$

we shall have for the variation

$$\delta F = \int_a^b k_1(\xi)\delta y(\xi)\, d\xi + 2a \int_a^b y(\xi)\delta y(\xi)\, d\xi$$
$$+ 2 \int_a^b \int_a^b k_2(t, \xi) y(t) \delta y(\xi)\, dt\, d\xi,$$

([1]) For the definition of non-regular functionals of degree n, cf. FRÉCHET, (31).

and for the first derivative

$$F'[y(t); \xi] = k_1(\xi) + 2ay(\xi) + 2\int_a^b k_2(t, \xi) y(t) dt,$$

which is a functional of the first degree of y, depending, however, *specially* (No. 19) on the value of y at the point ξ. In order to find the maxima and minima of a *regular* functional of the second degree, we have seen just above that we must equate to zero the first derivative, i.e. we have to solve the functional equation

$$k_1(\xi) = -2\int_a^b k_2(t, \xi) y(t) dt$$

(a linear integral equation of the first kind; cf. Chapter II) with respect to the unknown function $y(t)$. To find the maxima and minima of the above *non*-regular functional, on the other hand, we have to solve the equation

$$y(\xi) = -\frac{k_1(\xi)}{2a} - \frac{1}{a}\int_a^b k_2(t, \xi) y(t) dt$$

(a linear integral equation of the second kind).

36. For the functional

$$F[y(t)] = \int_a^b f(t, y(t), y'(t), \ldots y^{(n)}(t)) dt,$$

which appears in the calculus of variations, the variation is a linear functional of $\delta y(t)$ which depends *specially* on the values of $\delta y(t)$ and its derivatives to the $(n-1)$th order at the extremities of the interval (a, b) and is given by

$$\delta F = \int_a^b \left[\frac{\partial f}{\partial y} - \frac{d}{dt}\left(\frac{\partial f}{\partial y'}\right) + \frac{d^2}{dt^2}\left(\frac{\partial f}{\partial y''}\right) \cdots \right.$$
$$\left. \cdots + (-1)^n \frac{d^n}{dt^n}\left(\frac{\partial f}{\partial y^{(n)}}\right)\right] \delta y(t) dt$$

+ terms depending on the values of $\delta y, \delta y', \ldots \delta y^{(n-1)}$ at a and b. The functional is therefore *derivable* at any point ξ whatever *within* the interval (a, b), and the first derivative

OPERATIONS ON FUNCTIONALS

$$F'[y(t); \xi] = \frac{\partial f}{\partial y} - \frac{d}{d\xi}\left(\frac{\partial f}{\partial y'}\right) + \frac{d^2}{d\xi^2}\left(\frac{\partial f}{\partial y''}\right) \cdots$$

$$\cdots + (-1)^n \frac{d^n}{d\xi^n}\left(\frac{\partial f}{\partial y^{(n)}}\right)$$

is an *ordinary function* of ξ, $y(\xi)$, $y'(\xi)$, ... $y^{(n)}(\xi)$; and in order to find the functions $y(\xi)$ for which the functional is a maximum or minimum, we must, by the preceding section, solve the ordinary (EULER's) differential equation of degree $2n$

$$\frac{\partial f}{\partial y} - \frac{d}{d\xi}\left(\frac{\partial f}{\partial y'}\right) + \frac{d^2}{d\xi^2}\left(\frac{\partial f}{\partial y''}\right) \cdots + (-1)^n \frac{d^n}{d\xi^n}\left(\frac{\partial f}{\partial y^{(n)}}\right) = 0.$$

37. If we consider the functional

$$F[y(t)] = \int_a^b f_1[\xi, y(\xi), y'(\xi)]\,d\xi$$
$$+ \int_a^b \int_a^b f_2[\xi, \eta, y(\xi), y(\eta), y'(\xi), y'(\eta)]\,d\xi\,d\eta,$$

where f_2 is supposed symmetrical with respect to ξ and η whatever y may be, its variation will be given by the formula

$$\delta F = \int_a^b \delta y(\xi)\,d\xi \left[\frac{\partial f_1}{\partial y} - \frac{d}{d\xi}\left(\frac{\partial f_1}{\partial y'}\right) + 2\int_a^b \left\{\frac{\partial f_2}{\partial y(\xi)} - \frac{d}{d\xi}\left(\frac{\partial f_2}{\partial y'(\xi)}\right)\right\}d\eta\right]$$

+ terms depending on the values of δy at the extremities a and b.

Equating to zero the functional derivative $F'[y(t); \xi]$, defined for every value of ξ within the interval (a, b), we shall then have the functional equation with respect to the unknown y

$$\frac{\partial f_1}{\partial y} - \frac{d}{d\xi}\left(\frac{\partial f_1}{\partial y'}\right) + 2\int_a^b \left[\frac{\partial f_2}{\partial y(\xi)} - \frac{d}{d\xi}\left(\frac{\partial f_2}{\partial y'(\xi)}\right)\right]d\eta = 0,$$

a functional equation of a different kind from the previous ones; as it has the characters of both integral and differential equations, it will be called an *integro-differential* equation. The equations of this type include as particular cases both *integral equations* (containing no derivatives of the unknown function) and *differential equations* (in which the unknown function does not appear under the sign of integration).

§ 6. Integration of Functionals.

38. For functions of n variables, $f(y_1, y_2, \ldots y_n)$, we can define the integral I given by

$$I = \int_{a_1}^{b_1} dy_1 \int_{a_2}^{b_2} dy_2 \ldots \int_{a_n}^{b_n} f(y_1, y_2, \ldots y_n) \, dy_n$$

where the integration extends to a parallelopiped P_n in space of n dimensions of volume $V_n = (b_1 - a_1)(b_2 - a_2) \ldots (b_n - a_n)$; and this integral divided by V_n can serve to give an idea of the order of magnitude of the values of f in the complex (the quotient $\dfrac{I}{V_n}$ is the *mean* of the values of f in P_n). Similarly, we can try to define something analogous for functionals, by the usual method of passing from a finite to an infinite number of variables.

A first difficulty arises, however, when we try to define the *volume* or *measure* of a functional field or aggregate in such a way that it shall be an *additive function* of the aggregate. Thus, for instance, for the field of all the functions $y(t)$ for which $y_1(t) \leq y(t) \leq y_2(t)$, $y_2(t) - y_1(t) > 0$ (which might be called a parallelopiped of ∞ dimensions), the volume, that is, the product of all the possible differences $y_2(t) - y_1(t)$, would be 0 if $y_2(t) - y_1(t) < 1$, and infinite if $y_2(t) - y_1(t) > 1$.

39. Instead of defining separately the integral and the volume of a functional field, we shall try to define directly the *mean* of a functional in a certain field. Observing that in the case of the ordinary functions $f(y_1, y_2, \ldots y_n)$ the mean coincides with the integral if the field considered is the cube of n dimensions $0 \leq y_i \leq 1$ ($i = 1, 2, \ldots n$), and in this case takes a simpler form, we shall limit our problem to calculating the *mean* of a functional $\mathrm{F}[y(t)]\,{}_a^b$ defined for $y(t)$ always between 0 and 1.

Taking $y(t)$ in any way whatever in this field, let us consider $\mathrm{F}[y(t) + h] = f(h)$ as a function of h defined in the interval $(0, 1)$ (substituting $y(t) + h - 1$ for $y(t) + h$ for those values of t for which the latter quantity > 1), and let us calculate the integral

$$\mathrm{I}_1[y(t)] = \int_0^1 \mathrm{F}[y(t) + h] \, dh.$$

If we denote the upper and lower limits of the functional in the field by L and l respectively, the integral will also lie between these two limits, and, if $y(t)$ is fixed, will represent the *mean*, in the ordinary sense, of the values of F for the simple infinity of functions $y(t)+h$ (or $y(t)+h-1$ for those values of t for which $y(t)+h>1$).

Denoting by L_1, l_1, respectively the upper and lower limits of the integral $I_1[y(t)]$ when $y(t)$ varies in the usual field, we shall have $l \leq l_1 \leq L_1 \leq L$. Now let us divide the interval (a,b) into two parts (a,c) and (c,b); and taking $y(t)$ in any way whatever let us consider the function $\bar{y}(t, h_1, h_2)$ which $= y(t)+h_1$ in the interval (a,c), and $= y(t)+h_2$ in the interval (c,b) (in each case reduced by 1 for those values of t for which they are >1). The integral

$$I_2[y(t)] = \int_0^1 \int_0^1 F[\bar{y}(t, h_1, h_2)]\,dh_1\,dh_2$$

will represent the mean of the functional in the doubly infinite aggregate of the functions $y(t, h_1, h_2)$, which contains an infinite number of simply infinite aggregates of the preceding type, and will therefore always lie between l_1 and L_1.

If we denote by L_2 and l_2 respectively the upper and lower limits when $y(t)$ varies in any way in the field, we shall have

$$l \leq l_1 \leq l_2 \leq L_2 \leq L_1 \leq L.$$

If we divide each of the two intervals (a,c) and (c,b) in turn into two other partial intervals, and calculate the upper and lower limits L_3 and l_3 of the corresponding quadruple integral, and so on, it may happen that the non-decreasing sequence of the l_i's and the non-increasing sequence of the L_i's converge to a single limit I; this will be called the *mean* or the *integral* of the functional F in the field considered. It should however be noted that the convergence of the two sequences and the value of their limit *will depend in general on the method of division of (a,b) into partial intervals*.

This definition of an integral, which is the first in order of time, is due to GATEAUX [1].

---

[1] Cf. GATEAUX, (43), (45).

40. Another definition has been given by DANIELL ([1]), by considering functionals *of limited variation* defined in a functional field of an infinite and enumerable number of dimensions (functions of an infinite number of variables), and extending to these fields STIELTJES's concept of an integral; the process is analogous to what happens in fields of n dimensions when a function $\alpha(y_1, y_2, \ldots y_n)$ *of limited variation* is given, by means of which the STIELTJES integral,

$$\int\int \ldots \int f(y_1, y_2, \ldots y_n) \, dy_1 \, dy_2 \ldots dy_n \, \alpha(y_1, y_2, \ldots y_n),$$

is defined.

This integral of DANIELL has been applied to the study of Brownian movements by N. WIENER([2]).

([1]) Cf. DANIELL, (14).
([2]) Cf. WIENER, (91).

BIBLIOGRAPHY I.

(1) ARZELÀ (C.). — Funzioni di linee. (*R. Acc. dei Lincei. Rend.*, Vol. V, 1st half-year, 1889.)
(2) ARZELÀ (C.). — Sulle funzioni di linee. (*R. Acc. delle Scienze delle Istituto di Bologna. Mem.*, Series 5, Vol. V, 1895.)
(3) ARZELÀ (C.). — Sul principio di Dirichlet. (*R. Acc. Bologna*, Vol. I, 1896-97.)
(4) ARZELÀ (C.).—Sul'inversione di un sistema di funzioni. (*R. Acc. di Bologna*, Vol. VII, 1902-1903.)
(5) ARZELÀ (C.).— Sul limite di un integrale doppio. (*R. Acc. di Bologna*, 1907-1908.)
(6) ARZELÀ (C.).—Su alcune questioni di calcolo funzionale. (*R. Acc. delle Scienze delle Istituto di Bologna. Mem.*, Series VI, Vol. VII, 1910.)
(7) BAIRE (R.).—Leçons sur les fonctions discontinues. (Coll. Borel.) Gauthier-Villars, Paris, 1905.
(8) BOULIGAND (G.).—Sur les modes de continuité de certaines fonctionnelles. (*Bull. des Sciences Math.*, 2nd Series, Vol. XLVII, 1923.)
(9) COURANT (R.). — Über eine neue Klasse von kovarianten Funktionalausdrücken, welche aus Variationsproblemen entspringen. (*Gött. Nachr.* 1925.)
(10) DANIELE (E.).— Formole di derivazione funzionale. (*R. Acc. dei Lincei. Rend.*, Series 5, Vol. 24, 1st half-year, 1915.)
(11) DANIELE (E.)—Sulle derivate delle funzioni di linee inverse. (*Atti Acc. Gioenia. Catania*, Series 5, Vol. VIII, 1915.)
(12) DANIELL (P. J.).—A General Form of Integral. (*Annals of Mathematics*, Series 2, Vol. XIX, 1918.)
(13) DANIELL (P. J.).—Functions of limited variation in an infinite number of dimensions. (*Annals of Mathematics*, Vol. XXI, 1919.)
(14) DANIELL (P. J.).—Integrals in an infinite number of dimensions. (*Annals of Mathematics*, Vol. XX, 1919.)
(15) DANIELL (P. J.).—The Derivative of a Functional. (*Bull. of American Math. Soc.*, Series 2, Vol. XXV, New York, 1919.)
(16) DIENES (P.).—Versuch einer systematischen Begründung der Funktionalrechnung. (*Math. es termesz. Ungar-Budapest.* Band XXXIV, 1916.)
(17) EVANS (G. C.).—Topics from the theory and applications of functionals, including integral equations. (*American Math. Soc., Cambridge Colloquium*, 1916.)
(18) EVANS (G. C.).—An extension of Hadamard's formula for a linear functional. (*Bull. of American Math. Soc.*, Vol. 23, 1917.)

(19) FABRI (C.).—Sopra alcune proprietà generali delle funzioni che dipendono da altre funzioni. (*Atti. R. Acc. Torino*, Vol. XXV, 1890.)
(20) FABRI (C.).—Sopre le funzioni di iperspazii. (*Atti Inst. Veneto*, Series VII, Vol. IV, 1893.)
(21) FISCHER (C. A.).—Functions of surfaces with exceptional points of curves. (*American Journal of Math.*, Vol. 38, 1916.)
(22) FISCHER (C.| A.).—Note on the order of continuity of functions of lines. (*American Math. Soc. Bull.*, Vol. 23, 1916.)
(23) FISCHER (C. A.).—Linear Functionals of n-Space. (*American Math. Soc. Bull.*, Vol. 23, 1917. — *Annals of Math., Princeton Univ.*, Vol. XIX, 1917.)
(24) FISCHER (C. A.).—On bilinear and n-linear functionals. (*Annals of Math., Princeton Univ.*, Series 2, Vol. XIX. — *Proc. National Acad. of Sciences*, Vol. III, 1917.)
(25) FISCHER (C. A.).—Necessary and sufficient conditions that a linear transformation be completely continuous. (*American Math. Soc. Bull.*, Vol. 27, 1920-21.)
(26) FRÉCHET (M.).—Sur les opérations linéaires. (*American Math. Soc. Trans.*, Vol. V, 1904; Vol. VI, 1905.)
(27) FRÉCHET (M.).—Généralisation d'un théorème de Weierstrass. (*C. R. de l'Acad. des Sciences de Paris*, 2nd half-year, 1907.)
(28) FRÉCHET (M.).—Sur quelques points du calcul fonctionnel (*Thèse*). Paris, 1906. (*Rend. del Circ. Mat. di Palermo*, Vol. XXII, 2nd half-year, 1906.)
(29) FRÉCHET (M.).—Les fonctions d'une infinité de variables. (*Comp. rend. du Congrès des Soc. Sav.*, 1909.)
(30) FRÉCHET (M.).—Sur les fonctionnelles continues. (*Ann. de l'Ecole Normale sup.*, 3rd Series, Vol. 27, 1910.)
(31) FRÉCHET (M.).—Sur la notion de différentielle dans le calcul fonctionnel. (*Comp. rend. du Congrès des Soc. Sav.*, 1912.)
(32) FRÉCHET (M.).—Sur les fonctionnelles lineaires et l'intégrale de Stieltjes. (*Extrait des Comp. rend. du Congrès des Soc. Sav.*, 1913. Paris, 1914.)
(33) FRÉCHET (M.).—Sur la notion de différentielle d'une fonction de ligne. (*American Math. Soc. Trans.*, Vol. XV, 1914.)
(34) FRÉCHET (M.).—Sur l'integrale d'une fonctionnelle étendue a un ensemble abstrait. (*Soc. Math. de France, Bull.*, Vol. XLIII, 1915.)
(35) FRÉCHET (M.).—L'ecart de deux fonctions quelconques, (*C. R. de l'Acad. de Sciences de Paris*, Vol. 162, 1916.)
(36) FRÉCHET (M.).—Sur divers modes de convergence d'une suite de fonctions d'une variable. (*Bull. of the Calcutta Math. Soc.*, Vol. XI, 1924.)
(37) FRÉCHET (M.).—Prolongement des fonctionnelles continues sur un ensemble abstrait. (*Bull. des Sciences Math.*, 2nd Series, Vol. XLVIII, 1924.)
(38) GATEAUX (R.).—Sur la représentation des fonctionnelles continues. (*R. Acc. dei Lincei. Rend.*, Vol. XXII, 2nd half-year, 1913.)
(39) GATEAUX (R.).—Sur les fonctionelles continues et les fonctionnelles analytiques. (*C. R. de l'Acad. des Sciences de Paris*, Vol. 157, 1913.)

(40) GATEAUX (R.).—Sur la représentation des fonctionnelles continues. (*R. Acc. dei Lincei. Rend.*, Vol. XXIII, 1st half-year,1914.)
(41) GATEAUX (R.).—Représentation d'une fonctionnelle continue satisfaisant a la condition du cycle fermé. (*R. Acc. dei Lincei. Rend.*, Vol. XXIII, 1st half-year, 1914.)
(42) GATEAUX (R.).—Sur les fonctionelles d'ordre entier d'approximation. (*R. Acc. dei Lincei. Rend.*, Vol. XXIII, 1st half-year, 1914.)
(43) GATEAUX (R.).—Sur la notion d'intégrale dans le domaine fonctionnel et sur la théorie du potentiel; avec Note de Paul LÉVY. (*Soc. Math. de France. Bull.*, Vol. XLVII, 1919.)
(44) GATEAUX (R.).—Fonctions d'une infinité de variables indépendantes; avec Note de Paul LÉVY. (*Soc. Math. de France Bull.*, Vol. XLVII, 1919.)
(45) GATEAUX (R.).—Sur diverses questions du calcul fonctionnel. (*Soc. Math. de France. Bull.*, Vol. L, 1922.)
(46) HADAMARD (J.).—Sur les dérivées des fonctions de ligne. (*Soc. Math. de France. Bull.*, Vol. XXX, 1902.)
(47) HADAMARD (J.).—Sur les opérations fonctionnelles. (*C. R. de l'Acad. des Sciences de Paris*, Vol. 136, 1st half-year, 1903.)
(48) HADAMARD (J.).—Leçons sur le calcul des variations. Paris, 1910.
(49) HADAMARD (J.).—Le calcul fonctionnel. (*Enseig. Math.*, 14th Year, 1912.)
(50) HELLY (E.).—Über lineare Funktionaloperationen. (*Sitz. Ak. Wiss. in Wien*, Vol. 121, 1912.)
(51) HILDEBRAND (T. H.).—A contribution to the foundations of Fréchet's Calcul Fonctionnel. (*American Journal of Math.*, Vol. XXVI, 1912.)
(52) KAKEYA (S.).—On linear operations and some groups of continuous functions. (*Sc. Rep. Tohoku Imp. Uni. Sendai*, Vol. 4, 1915.)
(54) KAKEYA (S.).—Theorys of analytic functions of lines. (*Sc. Rep. Tohoku Imp. Uni. Sendai*, Vol. 6, 1917.)
(54) KAKEYA (S.).—On functions of lines and a set of curves. (*Sc. Rep. Tohoku Imp. Uni. Sendai*, Vol. 7, 1918.)
(55) LEBESGUE (H.).—Leçons sur l'intégration et la recherche des fonctions primitives. (Coll. Borel.) Gauthier-Villars, Paris, 1904.
(56) LE STOURGEON (E.).—Minima of functions of lines. (*American Math. Soc. Trans.*, Vol. 21, 1920.)
(57) LEVI (E. E.).—Funzioni di linee. (*Estrat. dal Boll. di Bibliografia e Storia delle Scienze Matematiche.* Paris, 1915.)
(58) LÉVY (P.).—Sur les dérivées des fonctions des lignes planes. (*C. R. de l'Acad. des Sciences de Paris*, Vol. 152, 1911.)
(59) LÉVY (P.).—Sur la notion de moyenne dans le domaine fonctionnel. (*C. R. de l'Acad. des Sciences de Paris*, Vol. 169, 1919.)
(60) LÉVY (P.).—Leçons d'analyse fonctionnel. Gauthier-Villars, Paris, 1922.
(61) LÉVY (P.).—Sur la dérivation et l'integration généralisées. (*Bull. des Sciences Math.* 2nd Series, Vol. XLVII, 1923.)
(62) LÉVY (P.).—Analyse fonctionnelle. (*Mémorial des Sciences Mathématiques*, 5th Fasc. Gauthier-Villars, Paris, 1925.)

(63) NALLI (P.).—Sulle operazioni funzionali lineari. (*Cir. mat. Palermo. Rend.*, Vol. 47, 1922.)
(64) PASCAL (E.).—Sui principi della teoria delle funzioni di linee, note I e II.—Gli integrali Riemanniani delle funzioni di linee, nota III.— Le formole di Green e di Stokes per le funzioni di linee, nota IV. (*R. Acc. di Napoli*, Series 3, Vol. XX, 1914.)
(65) PASCAL (E.).—Le linee funzioni di linee. (*Giornale di Mat. di Battaglini*, Vol. LIII. Naples, 1915.)
(66) PICONE (M.).—Le equazioni alle variazioni per cause pertubatrici variabili nel concetto di Volterra di variazione prima per una funzione di linea. (*R. Acc. dei Lincei. Rend.*, Vol. 28, 2nd half-year, 1919.)
(67) PICONE (M.).—Sulle funzioni additive di campo. (*R. Acc. dei Lincei. Rend.*, Vol. 28, Series 5, 1st half-year, 1919.)
(68) RADON (F.).—Theorie und Anwendungen der absolut additiven Mengenfunktionen. (*Ber. der K. Akad. der Wiss.*, Vol. CXII, Vienna, 1913.)
(69) RADON (F.).—Über lineare Funktionaltransformationen und Funktionalgleichungen. (*Wien Aug.* 56-189.—*Wien. Ber.* (2) 128, 1919.— *Sitz. K. Ak. in Wien*, Abt. 2, Vol. 128, 1919.)
(70) RASOR (S. E.).—On the integration of Volterra's derivatives. (*American Math. Soc. Bull.*, Vol. 22, 1916.)
(71) RIESZ (F.).—Stetigkeitsbegriff und abstrakte Mengenlehre. (*Atti* 4. *Congresso intern. dei Matematici*, Vol. II. Rome, 1909.)
(72) RIESZ (F.).—Sur les opérations fonctionnelles linéaires. (*C. R. de l'Acad. des Sciences de Paris*, 1909.)
(73) RIESZ (F.).—Sur certaines systèmes d'équations fonctionnelles et l'approximation des fonctions continues. (*C. R. de l'Acad. des Sciences de Paris*, Vol. 150, 1910.)
(74) RIESZ (F.).—Über lineare Funktionalgleichungen. (*Acta Math.*, Vol. 41, 1916.)
(75) STEINHAUS (A.).—Additive und stetige Funktionaloperationen. (*Math. Zeitschr.*, Vol. 5, 1919.)
(76) TONELLI (L.).—Sulle funzioni di linee. (*R. Acc. dei Lincei. Rend.*, Vol. XXIII, Series 5, 1st half-year, 1916.)
(77) TONELLI (L.).—La semicontinuità nel calcolo delle variazioni. (*Cir. mat. Palermo. Rend.*, Vol. 44, 1920.)
(78) TONELLI (L.).—Fondamenti di calcolo delle variazioni, Vols. I and II. Zanichelli, Bologna, 1921 and 1923.
(79) TRICOMI (F.).—Sull'iterazioni delle funzioni di linee. (*Giornale di Mat. di Battaglini*, Vol. 55 [(3) 8], 1917.
(80) TRICOMI (F.).—Sulle serie di funzioni di linee. (*Acc. Scienze di Napoli. Rend.*, Series 3, Vol. XXVI, 1920.)
(81) TRICOMI (F.).—Le serie di potenze nel campo delle funzioni di linee. (*Acc. Scienze di Napoli. Rend.*, Series 3, Vol. XXVI, 1920.)
(82) VESSIOT (E.).—Sur la théorie des multiplicités et le calcul des variations. (*Bull. Soc. Math. de France*, Vol. 40, 1912.)
(83) VÒLTERRA (V.).—Sopra le funzioni che dipendono da altre funzioni. (*R. Acc. dei Lincei. Rend.*, Vol. VI, 1887 [3 notes].)

I BIBLIOGRAPHY 39

(84) VOLTERRA (V.).—Sopra le funzioni dipendenti da linee. (*R. Acc. dei Lincei. Rend.*, Series 4, Vol. III, 2nd half-year, 1887 [2 notes].)

(85) VOLTERRA (V.).—Leçons sur l'intégration des équations différentielles aux dérivées partielles. Stockholm, 1906. (Upsala, 1906.)

(86) VOLTERRA (V.).—Trois leçons sur quelques progrès récents de la physique mathématique. (Lectures delivered at the celebration of the 20th anniversary of the foundation of Clark University. Worcester, Mass., 1909.)

(87) VOLTERRA (V.).—Leçons sur les fonctions de lignes. (Coll. Borel.) Gauthier-Villars, Paris, 1913.

(88) VOLTERRA (V).—Les problèmes qui ressortent du concept de fonction de ligne. (*Sitzungsber. Berliner Math. Gesell.*, XIII. Jahrgang, 6. Sitz., 1914.)

(89) VOLTERRA (V).—Saggi scientifici. (L'evoluzione delle idee fondamentali del calcolo infinitesimale.—L'applicazione del calcolo ai fenomeni d'eredità.) N. Zanichelli, Bologna, 1924.

(90) WEIERSTRASS (K.).—Über die analytische Darstellbarkeit sogenannter willkürlicher Funktionen reeller Argumente. *Math. Werke*, III. Bd. Berlin, 1903.

(91) WIENER (N.).—Differential space. (*Publications of the Massachusetts Institute of Technology.* Series II, No. 60, June 1923.)

(92) WINTER (M.).—Les principes du calcul fonctionnel. (*Rev. de Métaphysique et de Morale*, 21st year, 1913.)

(93) WISHNEWSKY (L. A.).—The absolute extremum of a certain polinomial functional. (*Proc. Labor Cauric-Univ.*, 1, 1920.)

CHAPTER II.

PROBLEMS OF THE FUNCTIONAL CALCULUS.
INTEGRAL EQUATIONS.

§ 1.

41. We have seen in the preceding chapter how from problems of the calculus of variations we pass naturally to equations in which the unknown is no longer, as in algebra, a number or a finite aggregate of numbers, but a function. If in fact we wish to determine a function $y(t)$ which shall make the value of the derivable functional $F[y\,\overset{b}{\underset{a}{(t)}}]$ (No. 27) a maximum or a minimum, then for this function $y(t)$ the variation δF of the functional itself must be zero, where δF is given by (Nos. 27 and 28)

$$\delta F = \int_a^b F'[y(t); x]\,\delta y(x)\,dx,$$

whatever $\delta y(x)$ may be; hence

$$F'[y(t); x] = 0, \tag{1}$$

i.e. the derivative must vanish identically, for every value of x.

We shall now show that the equations of this type are obtained as generalisations of ordinary systems of n equations in n unknowns

$$\left.\begin{aligned} f_1(y_1, y_2, \ldots y_n) &= z_1 \\ f_2(y_1, y_2, \ldots y_n) &= z_2 \\ \cdots\cdots\cdots\cdots\cdots\cdots & \\ f_n(y_1, y_2, \ldots y_n) &= z_n \end{aligned}\right\}, \tag{2}$$

or more shortly

$$f_i(y_1, y_2, \ldots y_n) = z_i, \qquad i = 1, 2, \ldots n. \tag{2'}$$

CHAP. II   THE FUNCTIONAL CALCULUS   41

Substituting continuous parameters x or t, variable between a and b, for the discontinuous indices, variable from 1 to n, the function $f_i(y_1, y_2, \ldots y_n)$ of the n variables y_k becomes a *functional* of a function $y(t)$, depending also on the continuous parameter x instead of on the index i; and the system (2) or (2′) is transformed into the more general *functional equation*

$$F[y(t); x] = z(x) \qquad (3)$$

in the unknown function $y(t)$. The problem of solving equation (3) therefore corresponds in the functional calculus to the inversion of the system (2) in the ordinary calculus, and can therefore be said to consist in the *inversion* of the functional

$$F[y(t); x] = z(x)$$

with respect to the unknown function $y(t)$. If we write (3) in the other form

$$F[y(t); x] - z(x) = G[y(t); x] = 0,$$

we see that these equations of type (3) include as a particular case the equations of type (1) which arise out of the calculus of variations.

A particular case of equations (3) is when $F[y(t); x]$ is a functional of degree m (No. 23); this case corresponds in ordinary algebra to a system (2) in which the functions f_i are polynomials of degree m.

§ 2.

42. A still more special case, but a very interesting one because, among other reasons, it serves as a base for the study of equations of higher degree, is when $F[y(t); x]$ is a *linear* functional of $y(t)$, corresponding to the case of a system (2) of n ordinary linear equations in n unknowns. If the linear functional is *regular* (Nos. 19, 20, 21), i.e. if it has no exceptional points, the equation to be considered will be

$$\int_a^b K(x, t) y(t) dt = z(x) \qquad (1)$$

(a linear integral equation *of the first kind*); if the functional has the exceptional point $t = x$, the equation will be of the type

$$\int_a^b K(x, t) y(t) \, dt + a_0 y(x) + a_1 y'(x) + \cdots + a_n y^{(n)}(x) = z(x) . \quad (2)$$

When $a_1 = a_2 = \cdots = a_n = 0$, $a_0 = 1$, the equation reduces to

$$\int_a^b K(x, t) y(t) \, dt + y(x) = z(x) \quad (3)$$

(FREDHOLM's integral equation *of the second kind*); if instead of this $a_1 = a_2 = \cdots = a_n = 0$, $a_0 = a(x)$, we have an equation

$$\int_a^b K(x, t) y(t) \, dt + a(x) y(x) = z(x), \quad (4)$$

which PICARD has called an equation *of the third kind*. If there are several exceptional terms in (2), the equation itself, containing the unknown function both under the integral sign and under the symbol of differentiation, must be classified among those we have called ordinary *integro-differential equations* (cf. Chapter V).

If the linear functional $F[y(t); x]$ depends on all the possible values of $y(t)$ when t varies between a and x ($a \leq x \leq b$), we shall have correspondingly the same number of other equations (called *Volterra's equations*, he having been the first to study them exhaustively)

$$\int_a^x K(x, t) y(t) \, dt = z(x) \quad \text{(linear equation of the first kind)}, \quad (1')$$

$$\int_a^x K(x, t) y(t) \, dt + y(x) = z(x) \text{ (linear equation of the second kind)}, (3')$$

etc. These VOLTERRA's equations, in which the upper limit of the integral is variable with the parameter x, do not differ substantially from the preceding equations in which the upper limit is constant ($= b$). In fact, if we admit that the function $K(x, t)$, which is called the *kernel* of the equation, can be discontinuous (with discontinuities of the first kind), we need only suppose that in the equations (1), (2), (3), (4), with constant limits, $K(x, t) = 0$ for $t > x$, in order that they may reduce to the corresponding VOLTERRA's equations with a variable limit.

§ 3. *Volterra's Equations of the Second Kind.*

43. We shall confine our attention for the moment to the four following fundamental types of linear integral equations:

$$\left. \begin{array}{l} y(x) + \int_0^x K(x,t) y(t) dt = z(x) \\ \int_0^x K(x,t) y(t) dt = z(x) \end{array} \right\} \text{(upper limit variable)},$$

$$\left. \begin{array}{l} y(x) + \lambda \int_0^1 K(x,t) y(t) dt = z(x) \\ \int_0^1 K(x,t) y(t) dt = z(x) \end{array} \right\} \text{(upper limit constant)}.$$

To solve these ([1]) we shall start from the methods in use for the solution of ordinary systems of n linear equations in n unknowns, and shall then, with the usual procedure, pass from a finite number to a continuous infinity of unknowns; we shall thus finally have the general solution of our functional equations.

We shall start with VOLTERRA's integral equation of the second kind

$$y(x) + \int_0^x K(x,t) y(t) dt = z(x) \qquad (1)$$

in which we shall suppose both K and z finite continuous functions of their arguments. This equation can be regarded as generated by a system of n linear equations in n unknowns of the type

$$y_i + \sum_{r=1}^{i-1} K_{ir} y_r = z_i \qquad (i = 1, 2, \ldots n) \qquad (2)$$

when the discontinuous index i is replaced by the continuous variable x, and r by t. These linear systems (2) have however a peculiarity which simplifies their solution very much. In fact, if we denote by Δ the determinant of the coefficients, and by Δ_{ir} the algebraic complement (or minor) of K_{ir} in Δ, we shall have

---

([1]) Cf. VOLTERRA, (76)—(82); cf. also (86)—(88) of Bibliography I.

THEORY OF FUNCTIONALS

$$\varDelta = \begin{vmatrix} 1 & 0 & 0 & \ldots & 0 \\ K_{21} & 1 & 0 & \ldots & 0 \\ K_{31} & K_{32} & 1 & \ldots & 0 \\ \cdot & \cdot & \cdot & & \cdot \\ \cdot & \cdot & \cdot & & \cdot \\ K_{n1} & K_{n2} & K_{n3} & \ldots & 1 \end{vmatrix} = 1, \qquad (3)$$

and therefore for the unknowns y_r, applying CRAMER's rule, and observing that $\varDelta_{ir} = 0$ for $i > r$, and $\varDelta_{rr} = 1$, we shall have simply

$$y_r = z_r + \sum_{1}^{r-1} \varDelta_{ir} z_i = z_r + \sum_{1}^{r-1} S_{ri} z_i \qquad (4)$$

if $\varDelta_{ir} = S_{ri}$.

Working out a similar procedure for equation (1), which we shall see better when we consider the equation of the second kind with constant limits, we should get the solution $y(x)$ we are in search of in the form

$$y(x) = z(x) + \int_0^x S(x, t) z(t) dt, \qquad (5)$$

where $S(x, t)$ is a suitable function which is called the *reciprocal* or *solvent kernel* of (1).

To obtain the determinants $\varDelta_{ir} = S_{ri}$, we put the solution (4) into the equation (2). We have

$$z_i + \sum_{1}^{i-1} S_{is} z_s + \sum_{1}^{i-1} z_s \left(K_{is} + \sum_{s+1}^{i-1} K_{ir} S_{rs} \right) = z_i.$$

Therefore

$$S_{is} + K_{is} + \sum_{s+1}^{i-1} K_{ir} S_{rs} = 0. \qquad (6)$$

Let us put

$$S_{is} = K_{is}^{(1)} + K_{is}^{(2)} + \cdots + K_{is}^{(i-s)},$$

where $K_{is}^{(1)}, K_{is}^{(2)}, \ldots K_{is}^{(i-s)}$ are the terms of degree $1, 2, \ldots i-s$. We then get from the preceding equation

$$\left. \begin{aligned} K_{is}^{(1)} &= -K_{is} \\ K_{is}^{(h)} &= \sum_{s+1}^{i-1} K_{ir}^{(1)} K_{rs}^{(h-1)} \end{aligned} \right\}. \qquad (7)$$

THE FUNCTIONAL CALCULUS

We thus have a rule for calculating the determinants S_{is}. From this rule we can deduce a method of calculating $S(z,t)$ by passing from the finite to the infinite, and we obtain

$$K^{(1)}(x,t) = -K(x,t)$$
$$K^{(2)}(x,t) = \int_t^x K^{(1)}(x,\xi) K^{(1)}(\xi,t) d\xi$$
$$K^{(3)}(x,t) = \int_t^x K^{(1)}(x,\xi) K^{(2)}(\xi,t) d\xi \qquad (8)$$
$$\cdots\cdots\cdots\cdots\cdots\cdots\cdots\cdots$$
$$K^{(h)}(x,t) = \int_t^x K^{(g)}(x,\xi) K^{(h-g)}(\xi,t) d\xi$$

(the *iterative kernels* of the kernel $K^{(1)}$), and

$$S(x,t) = \sum_1^\infty K^{(h)}(x,t). \qquad (9)$$

Proceeding more rigorously, $K(x,t)$ being limited in absolute value (i.e. for any x and t whatever, $|K(x,t)| \leq M$, and therefore also $|K^{(1)}(x,t)| \leq M$), we shall have for (8)

$$|K^{(2)}(x,t)| \leq M^2 |x-t|$$
$$|K^{(3)}(x,t)| \leq \frac{M^3 |x-t|^2}{2!}$$
$$\cdots\cdots\cdots\cdots\cdots\cdots$$
$$|K^{(h)}(x,t)| \leq \frac{M^h |x-t|^{h-1}}{(h-1)!}.$$

The series (9) is therefore always uniformly convergent (principle of convergence), and

$$\lim_{h=\infty} K^{(h)}(x,t) = 0. \qquad (10)$$

We shall have also (principle of reciprocity) (cf. (6))

$$\int_x^t K(x,\xi) S(\xi,t) d\xi = \int_x^t S(x,\xi) K(\xi,t) d\xi = K(x,t) + S(x,t), \qquad (11)$$

and therefore, multiplying both sides of (1) (with ξ put instead of x) by $S(x,\xi)$, and integrating with respect to ξ from 0 to x,

$$\int_0^x S(x,\xi) y(\xi) d\xi + \int_0^x S(x,\xi) d\xi \int_0^\xi K(\xi,t) y(t) dt = \int_0^x S(x,\xi) z(\xi) d\xi,$$

46   THEORY OF FUNCTIONALS   CHAP.

and by DIRICHLET's rule and the preceding formula,

$$\int_0^x S(x,\xi)y(\xi)d\xi - \int_0^x K(x,t)y(t)dt - \int_0^x S(x,t)y(t)dt$$
$$= \int_0^x S(x,\xi)z(\xi)d\xi,$$

and finally, by (1) itself

$$y(x) - z(x) = \int_0^x S(x,t)z(t)dt,$$

or

$$y(x) = z(x) + \int_0^x S(x,t)z(t)dt.$$

a *formula which solves equation* (1) (principle of inversion).

The solution is evidently unique, since if there were another, $y_1(x)$, the difference $y_1(x) - y(x) = u(x)$ should satisfy the (homogeneous) equation

$$u(x) + \int_0^x K(x,t)u(t)dt = 0,$$

or

$$u(x) = \int_0^x K^{(1)}(x,t)u(t)dt,$$

and therefore also

$$u(x) = \int_0^x K^{(i)}(x,t)u(t)dt,$$

whatever i may be; but, by (10), the limit of the integral on the right is 0 for $i = \infty$, and therefore also $u(x) = 0$.

On constructing the iterative kernels $S^{(i)}(x,t)$ of $S^{(1)}(x,t) = -S(x,t)$, by the same method as that applied to the equations (8), it would be seen that the series $\Sigma_i S^{(i)}(x,t)$ is uniformly convergent and its sum coincides with $K(x,t)$, since it must, like $K(x,t)$, be a solution (the only one) of the integral equation (11) (principle of reciprocity).

The function $S(x,t)$ given by (9), which we have called the *reciprocal* or *solvent kernel* of the equation (1), is evidently a *functional* $\mathrm{F}[K(\xi_1, \xi_2); x, t]$ of the function of two variables $K(\xi_1, \xi_2)$; and if to it, regarded as a function of x und t, we again apply the same functional F, we shall again obtain, by the foregoing observation, the function $K(x, t)$; i.e. we shall have

$$F[S(\xi_1, \xi_2); x, t] = K(x, t).$$

We have thus an interesting example of a functional that coincides with its *inverse* functional (No. 41).

We may point out, lastly, that some of the hypotheses we have made on the nature of the functions $K(x, t)$ and $z(x)$ can be considerably widened (discontinuous kernels, etc.) ([1]).

We can solve the equation (1) by the method of *successive approximations*([2]). Applying this method, we can write

$$\left.\begin{aligned} y(x) &= z(x) - \int_0^x K(x,t) y(t) \, dt \\ y(x) &= z(x) - \int_0^x K(x,t) z(t) \, dt \\ &\quad + \int_0^x K(x,t) \, dt \int_0^t K(t,\xi) y(\xi) \, d\xi \end{aligned}\right\}, \quad (12)$$

and so on, always replacing y by its expression given by the first line of (12); and if in the expression thus obtained we make a suitable change in the order of integration, applying DIRICHLET's rule, we shall have, after $i - 1$ substitutions,

$$y(x) = z(x) + \int_0^x K^{(1)}(x,t) z(t) \, dt + \int_0^x K^{(2)}(x,t) z(t) \, dt + \cdots$$
$$+ \int_0^x K^{(i-1)}(x,t) z(t) \, dt + \int_0^x K^{(i)}(x,t) y(t) \, dt,$$
$$= z(x) + \int_0^x (K^{(1)}(x,t) + K^{(2)}(x,t) + \cdots$$
$$+ K^{(i-1)}(x,t)) z(t) \, dt + \int_0^x K^{(i)}(x,t) y(t) \, dt,$$

and if as i increases the last term tends to 0, we shall have $y(x)$ in the form (5).

44. The formulae (8), which define the iterative kernels $K^{(h)}(x, t)$ of the kernel $K^{(1)}(x, t)$ in terms of $K^{(g)}$ and $K^{(h-g)}$, give us a first example of a particular functional depending on two

---

([1]) Cf. § 8 below.
([2]) Cf. LE ROUX, (36); PICARD, (45); BÔCHER, (6); LALESCO, (33).

functions $K^{(g)}$ and $K^{(h-g)}$, each of two variables, which we shall study at greater length in Chapters IV and V (Section 1, § 3); it is called the *product by composition* of the first kind of the two functions $K^{(g)}$ and $K^{(h-g)}$, and is denoted by $\overset{*}{K}{}^{(g)}\overset{*}{K}{}^{(h-g)} = K^{(h)}$. If further we construct the product by composition of a function K by itself, we have a *power by composition* $\overset{*}{K}\overset{*}{K} = \overset{*}{K}{}^2 = K^{(2)}$, and in general $\overset{*}{K}{}^i = \overset{*}{K}{}^{i-1}\overset{*}{K} = (-1)^i K^{(i)}$; i.e. the ith iterative kernel $K^{(i)}$ is the ith power by composition of K; and if we construct a convergent series $\sum_{1}^{\infty} a_i \overset{*}{K}{}^i$, we have an expression which is called a *function by composition* of K; an example of this is the solvent kernel $S(x,t) = \sum_{1}^{\infty}(-1)^i \overset{*}{K}{}^i(x,t)$.

§ 4. *Fredholm's Equations of the Second Kind.*

45. The two methods of successive approximation and extension of CRAMER's formula from the case of a finite to an infinite number of variables can also be applied to integral equations of the second kind with constant limits

$$y(x) + \lambda \int_0^b K(x,t) y(t) dt = z(x). \tag{1}$$

By the method of successive approximations we proceed in the same way as that used in § 3, but this works only if λ is sufficiently small; in fact, if we construct the expressions, analogous to (8) in No. 43,

$$\left. \begin{array}{l} K^{(1)}(x,t) = - K(x,t) \\ K^{(2)}(x,t) = \int_0^b K^{(1)}(x,\xi) K^{(1)}(\xi,t) d\xi \\ \cdots\cdots\cdots\cdots\cdots\cdots\cdots\cdots\cdots \\ K^{(i)}(x,t) = \int_0^b K^{(i-1)}(x,\xi) K^{(1)}(\xi,t) d\xi \end{array} \right\}, \tag{2}$$

the *iterative kernels* or *powers by composition* of $K^{(1)}$ *of the second kind* (cf. Chapter IV, § 2), and if we suppose $|K^{(1)}(x,t)| \leq M$, we shall have for $K^{(i)}(x,t)$ the limitation

$$|K^{(i)}(x,t)| \leq M^i b^{i-1},$$

and therefore the series

$$S(x,t;\lambda) = \sum_i \lambda^{i-1} K^{(i)}(x,t) \qquad (3)$$

will certainly be convergent for $|\lambda M b| < 1$, i.e. for $|\lambda| < \frac{1}{bM}$; but it is not known whether it will be so for $|\lambda| \geq \frac{1}{bM}$. When the series (3) is uniformly convergent for t variable in the interval $(0, b)$, the solution of (1) will be unique, and will be given by

$$y(x) = z(x) + \lambda \int_0^b S(x,t;\lambda) z(t) dt. \qquad (4)$$

The usual method of passing from a finite number to a continuous infinity of variables already applied in the case of VOLTERRA's equation is, however, much more satisfactory. If in fact we start from the linear system

$$y_i + \lambda \sum_1^n k_{ir} y_r = z_i \quad (i = 1, 2, \ldots n), \qquad (5)$$

which corresponds to the equation (1) in the case of a finite number n of unknowns y_r ($r = 1, 2, \ldots n$), it may be pointed out that the (unique) solution of the system is obtained, by CRAMER's rule, when the determinant $\Delta(\lambda)$ of the coefficients is different from zero, in the form

$$y_r = \frac{\sum_i^n \Delta_{ir}(\lambda) z_i}{\Delta(\lambda)} = z_r + \lambda \frac{\sum_i^n \Delta'_{ir}(\lambda) z_i}{\Delta(\lambda)}, \qquad (6)$$

where we have denoted by $\Delta_{ir}(\lambda)$ (a polynomial of degree $n-1$ in λ) the algebraic complements of the determinant $\Delta(\lambda)$ (a polynomial of degree n in λ), and by $\Delta'_{ir}(\lambda)$ the same expressions $\Delta_{ir}(\lambda)$ divided by λ for $i \neq r$, and the polynomial $\frac{\Delta_{ir}(\lambda) - \Delta(\lambda)}{\lambda}$ for $i = r$. Passing from a finite to an infinite number of unknowns, we can show, analogously [1], that in the case of equation (1) it is possible to construct two integral functions in λ (instead of polynomials), $D_k(\lambda)$ and $D_k(x,t;\lambda)$, functionals of the ker-

---

[1] Cf. FREDHOLM, (20).

nel $K(x, t)$, of which the first corresponds to the determinant $\Delta(\lambda)$ of the unknowns, and the second, depending also on the parameters x and t, to the quantities $\Delta'_{ir}(\lambda)$; so that if we put

$$S(x, t; \lambda) = -\frac{D_k(x, t; \lambda)}{D_k(\lambda)}$$

when $D_k(\lambda)$ is different from zero (S is a *meromorphic* function of λ), the solution, which is unique, of the equation (1) is given by the formula (4), corresponding exactly to (6).

If now we consider those values λ_i of λ for which the *determinant* $D_k(\lambda)$ of the integral equation vanishes (characteristic values), it can be shown that, analogously to what happens for the system (6), the *homogeneous* integral equation obtained from (1) by putting $z(x) = 0$ has, similarly, solutions $\varphi_i(x)$ which are not identically zero (*characteristic functions*), and therefore equation (6), *when it is possible*, has not one but an infinite number of solutions. For further details on this extremely interesting chapter in the theory of functionals the reader is referred to the numerous publications on the subject [1].

§ 5. Symmetrical Kernels; Regular Homogeneous Functionals of the Second Degree.

46. A particularly interesting group of integral equations is that for which the kernel $K(x, t)$ is a symmetrical function of the two variables x and t, corresponding to linear systems (5) in which $k_{ir} = k_{ri}$. For these symmetrical kernels it has been proved that *characteristic values* λ_i, *all real*, do effectively exist, or, in other words, that the functional $D_k(\lambda)$ (considered as a function of λ) vanishes at points of the real axis [2]. This property is the natural extension to the case of an infinite number of variables of the well-known properties of the so-called *secular equations*. Much attention has therefore also been given to those kernels which, while not symmetrical, are yet capable of being made so (*symmetrizable*)[3].

---

[1]. Cf. VIVANTI, (73); HILBERT, (23); BÔCHER, (6); LALESCO, (33).
[2] Cf. HILBERT, (23).
[3] Cf. GOURSAT, (21), Vol. III, p. 466.

A symmetrical kernel $K(x, t)$ can be put in the form of a uniformly convergent series

$$K(x, t) = - \sum_{1}^{\infty}{}_i \frac{1}{\lambda_i} \varphi_i(x) \varphi_i(t), \qquad (1)$$

where $\varphi_1, \varphi_2, \ldots \varphi_i, \ldots$ are the *normal* system ([1]) of characteristic functions, corresponding respectively to the characteristic values λ_i for which

$$\varphi_i(x) + \lambda_i \int_0^b K(x, t) \varphi_i(t) dt = 0.$$

In this case the regular homogeneous functional of the second degree

$$F[y(t)] = \int_0^b \int_0^b K(x, t) y(x) y(t) dx dt,$$

in which $K(x, t)$ can always be supposed symmetrical (analogous to a form of the second degree in n variables), can be put in the *canonical form* ([2])

$$F[y(t)] = - \sum_{1}^{\infty}{}_i \frac{1}{\lambda_i} \Big(\int_0^b \varphi_i(t) y(t) dt \Big)^2,$$

an expression which corresponds perfectly to the canonical form of a form of the second degree. These properties can be extended to unsymmetrical kernels: this has been done by SCHMIDT ([3]).

§ 6. *Integral Equations of the First Kind with Constant Limits.*

47. Given an integral equation of the first kind with constant limits

$$z(x) = \int_0^b K(x, t) y(t) dt, \qquad (1)$$

we can always suppose the kernel K symmetrical. For if not,

([1]) Cf. HILBERT, (23); GOURSAT, (21).
([2]) Cf. HILBERT, (23).
([3]) Cf. SCHMIDT, (60).

multiplying (1) by $K(x, \xi)$ and integrating with respect to x, we should have

$$\int_0^b K(x, \xi) z(x) dx = \int_0^b \int_0^b K(x, \xi) K(x, t) y(t) dx dt,$$

and putting

$$\bar{z}(\xi) = \int_0^b K(x, \xi) z(x) dx, \qquad H(\xi, t) = \int_0^b K(x, t) K(x, \xi) dx,$$

(H a symmetrical function of ξ and t), we should obtain from (1) the other equation of the same type

$$\bar{z}(\xi) = \int_0^b H(\xi, t) y(t) dt.$$

If then the symmetrical kernel $K(x, t)$ of (1) is closed [1], i.e. if there is no function $\Psi(x)$ for which

$$\int_0^b K(x, t) \Psi(x) dx = 0$$

identically for any value of t whatever, we can try to expand the unknown function $y(t)$ as a series in terms of the characteristic functions $\varphi_i(t)$,

$$y(t) = \sum_1^\infty a_i \varphi_i(t), \qquad (2)$$

in which, the functions φ_i being normal among themselves, the coefficients a_i would be given by

$$a_i = \int_0^b \varphi_i(t) y(t) dt.$$

But then, adopting for $K(x, t)$ the expansion (1) of No. 46, we get from (1)

$$\int_0^b z(x) \varphi_i(x) dx = - \sum_1^\infty \frac{1}{\lambda_r} \Big(\int_0^b \varphi_r(x) \varphi_i(x) dx \Big) \Big(\int_0^b \varphi_r(t) y(t) dt \Big),$$

and from this, putting $c_i = \int_0^b z(x) \varphi_i(x) dx$, and remembering

---

[1] Cf. GOURSAT, (21).

that the functions φ_r form a normal system, we get

$$c_i = -\frac{1}{\lambda_i} a_i, \qquad a_i = -\lambda_i c_i. \qquad (3)$$

Hence if the solution $y(t)$ of (1) exists and can be expanded in the series (2), the coefficients will be given by (3). This is the method given by LAURICELLA in various writings of his ([1]).

PICARD and others ([2]) have shown further that the *necessary and sufficient condition* for the existence of the solution of equation (1) is that the series $\sum_{1}^{\infty} \lambda_i^2 c_i^2$ should be convergent, and in this case the solution is given by (2) and (3).

A simpler method of solving equation (1) can be applied in the case when both the kernel $K(x, t)$ and the known function $z(x)$ are analytical functions of their arguments, and the kernel $K(x, t)$ has a *pole of the first order* for $t = s(x)$ (its position varying as x varies), with residue $r(x)$ ([3]). In this case, if the solution $y(t)$ exists and is regular, and is not equal to 0, we have that $z(x)$ must be polydromic and have for critical points those points β for which $s(\beta) = b$ or 0 (one of the extremities of the integral); and denoting by $z_0(x)$ and $z_1(x)$ the values of $z(x)$ on two consecutive branches starting from the point β, and by $\bar{s}(t) = x$ the inverse function of $t = s(x)$, the solution itself, if it exists, is given by the simple formula

$$y(t) = \frac{1}{2\pi i r \dot{s}(t)} \left\{ z_1(s(t)) - z_0(\bar{s}(t)) \right\}.$$

In the case of a pole not of the first order, but of order n, the determination of the solution $y(t)$, when it exists, comes back to the *integration of a linear differential equation in y of order $n-1$*.

§ 7. *Volterra's Integral Equation of the First Kind.*

48. Passing to the study of VOLTERRA's integral equation of the first kind

$$z(x) = \int_0^x K(x, t) y(t) \, dt, \qquad (1)$$

([1]) Cf. LAURICELLA, (34) and (35).
([2]) Cf. PICARD, (47) and (48).
([3]) Cf. FANTAPPIÈ, (19).

where $z(0) = 0$, it may be pointed out that, with suitable hypotheses as to regularity for the kernel K and the function $z(x)$, we have on differentiating once

$$z'(x) = K(x,x)y(x) + \int_0^x \frac{\partial K(x,t)}{\partial x} y(t) dt, \qquad (2)$$

and, if $K(x,x) \neq 0$, we reach the equation of the second kind

$$\frac{z'(x)}{K(x,x)} = y(x) + \int_0^x \frac{\frac{\partial K}{\partial x}}{K(x,x)} y(t) dt \qquad (3)$$

already studied.

Instead of differentiating (1), we could integrate by parts and so reach the other equation

$$z(x) = K(x,x)\theta(x) - \int_0^x \frac{\partial K(x,t)}{\partial t} \theta(t) dt, \qquad (4)$$

in which we have put

$$\theta(x) = \int_0^x y(t) dt,$$

and therefore also, if $K(x,x) \neq 0$, the equation of the *second kind* with respect to the unknown $\theta(x)$

$$\frac{z(x)}{K(x,x)} = \theta(x) - \int_0^x \frac{\frac{\partial K}{\partial t}}{K(x,x)} \theta(t) dt, \qquad (5)$$

from which we could find $\theta(x)$ and from it, by differentiating, $y(x) = \theta'(x)$.

§ 8. *Singular Kernels.*

49. *Case in which the kernel becomes infinite. Derivatives with any index whatever.* The above method breaks down, however, if $K(x,x) = 0$, or if $K(x,t)$ becomes singular for $t = x$, or, more generally, if the kernels of the equations (3) and (5) of No. 48 have singularities. It is rather curious to note that it was precisely these singular cases that were the first in order of time to arise; for example, the first integral equation considered

goes back to ABEL ([1]), and is as follows:

$$\sqrt{2g}\,z(x) = \int_0^x \frac{y(t)\,dt}{\sqrt{x-t}},$$

and the kernel $\dfrac{1}{\sqrt{x-t}}$ becomes infinite for $x=t$. This equation expresses the problem (a generalisation of that of the tautochronous curve) of determining a curve in a vertical plane such that a heavy particle obliged to move on it and starting with no velocity shall take the time $z(x)$ to descend a height x (the difference of level between the points of departure and arrival). For it, and for the more general equation

$$z(x) = \int_0^x \frac{y(t)\,dt}{(x-t)^\alpha} \quad (0 < \alpha < 1),$$

ABEL gave the solution

$$y(t) = \frac{\sin \alpha \pi}{\pi} \frac{d}{dt} \int_0^t \frac{z(x)\,dx}{(t-x)^{1-\alpha}},$$

a solution which we shall find below (in No. 50) by a general method of procedure.

Later on LIOUVILLE ([2]) too was led by his considerations on derivatives with any index α whatever to solve ABEL's integral equation over again without knowing it.

This study of derivatives with any index whatever was subsequently pursued by RIEMANN ([3]), H. HOLMGREN ([4]), and others, taking for definition

$$D_x^\alpha y(x) = \frac{1}{\Gamma(-\alpha)} \int_0^x (x-t)^{-\alpha-1} y(t)\,dt \quad \text{for} \quad \alpha < 0,$$

$$D_x^\alpha y(x) = \frac{d^p}{d\,x^p} D_x^{\alpha-p} y(x), \text{ for } 0 < \alpha < p \text{ (p an integral number)},$$

or equivalent formulae ([5]); these show clearly how *Volterra's integral equations* can easily be reached from this point of view([6]).

([1]) Cf. ABEL, (1) and (2).
([2]) Cf. LIOUVILLE, (40), (41), (42).
([3]) Cf. RIEMANN, (53).
([4]) Cf. H. HOLMGREN, (25).
([5]) Cf. also HADAMARD, (22), chap. V.
([6]) Cf. SCATIZZI, (59).

They have also been used by Dr. MANDELBROJT ([1]) in an elegant extension of the ordinary calculus of variations. In this calculus the quantity to be made maximum or minimum is an integral

$$I = \int_0^b f(x, y, y', \ldots y^{(n)}) \, dx$$

which relates to an ordinary function of x and of the $n+1$ variables $y^{(i)}$ ($i = 0, 1, \ldots n$, denoting differentiation). Dr. MANDELBROJT, however, applying the usual method of passing from a finite number to a continuous infinity of variables, considers an integral relating to a functional

$$F_\alpha [y^{(\alpha)}(x); x]$$

which depends on x and on all the values of the function $y^\alpha(x) = D_x^\alpha y(x)$ for α variable in an interval $(0, A)$. In order to make the quantity

$$I = \int_0^1 F_\alpha [y^{(\alpha)}(x); x] \, dx$$

maximum or minimum, we make the first variation δI vanish and apply DIRICHLET's rule instead of the ordinary integration by parts; we thus reach the result that the function $y(x)$ must satisfy a functional equation of the type

$$G_\alpha [y^{(\alpha)}(x); x] = 0, \qquad 0 \leq \alpha \leq A, \qquad y^{(\alpha)}(x) = D_x^\alpha y(x),$$

an equation which might be called an *ordinary continuous differential equation* in the unknown function $y(x)$.

More general integral equations than those of ABEL and LIOUVILLE have been discussed by SONINE ([2]), in which it is supposed that the nucleus $K(x, t) = K(x - t)$, a function of the difference $x - t$; but for all these particular cases of integral equations, treated by various and laborious methods, there was no comprehensive theory until the general theory of linear integral equations came into being.

50. *Method of transformation of the kernel* ([3]). We shall next discuss the case of a VOLTERRA's equation of the first kind in

---

([1]) Cf. MANDELBROJT, (43).
([2]) Cf. SONINE, (61) and (62).
([3]) Cf. VOLTERRA, (76), Note 2.

which the kernel $K(x, t)$ becomes infinite of order $\alpha < 1$ for $x = t$; it can therefore be written

$$K(x, t) = \frac{G(x, t)}{(x - t)^\alpha}$$

and the integral equation will be

$$z(x) = \int_0^x \frac{G(x, t)}{(x - t)^\alpha} y(t)\, dt, \tag{1}$$

in which we shall further suppose both $G(x, t)$ and $\dfrac{\partial G}{\partial x}$ continuous and limited. To solve equation (1), we multiply both sides by $\dfrac{dx}{(\xi - x)^{1-\alpha}}$ and integrate from 0 to ξ; we shall have

$$\int_0^\xi \frac{z(x)\, dx}{(\xi - x)^{1-\alpha}} = \int_0^\xi \frac{dx}{(\xi - x)^{1-\alpha}} \int_0^x \frac{G(x, t)}{(x - t)^\alpha} y(t)\, dt.$$

Putting

$$\int_0^\xi \frac{z(x)\, dx}{(\xi - x)^{1-\alpha}} = f(\xi) \qquad (f(0) = 0),$$

$$\int_t^\xi \frac{G(x, t)\, dx}{(\xi - x)^{1-\alpha}(x - t)^\alpha} = L(\xi, t),$$

we shall have, by DIRICHLET's rule, the other integral equation of the first kind in the unknown function $y(t)$

$$f(\xi) = \int_0^\xi L(\xi, t) y(t)\, dt, \tag{2}$$

in which, however, *the transformed kernel $L(\xi, t)$ is finite.*
  In fact, putting

$$x = t + (\xi - t) u,$$

we have

$$L(\xi, t) = \int_0^1 \frac{G[t + (\xi - t)u, t]}{(1 - u)^{1-\alpha} u^\alpha}\, du,$$

and therefore, as we have supposed G limited, $|G| < M$, and we obtain

$$L(\xi,t) < M \int_0^1 \frac{du}{(1-u)^{1-\alpha} u^\alpha} = M \frac{\pi}{\sin a\pi}.$$

Having thus transformed equation (1) into (2), we can apply to the latter, which has its nucleus $L(\xi, t)$ regular, the method set out above, since it can be shown that every solution of (2) is also a solution of (1).

VOLTERRA integral equations of the second kind have been considered by EVANS in the case where the kernel becomes infinite to a non-integrable order. The kernel is taken in the form $A(x, t)/f(t)$, where $f(t)$ vanishes at $t = 0$ in such a way that $\int \frac{dt}{f(t)}$ need not be convergent; in this case there may be an infinite number of continuous solutions vanishing at $t = 0$. By means of a transformation of variables, the VOLTERRA equation with an infinite limit in the integral may be considered from this point of view; here with a bounded continuous kernel there may in special situations exist an infinite number of continuous solutions. EVANS gives also in this latter case sufficient conditions that there may exist a solvent kernel and a unique bounded solution. For this, the principal condition is that there shall be a constant $N < 1$ such that

$$\int_x^\infty |K(x,t)|\,dt < N < 1,$$

for x sufficiently large[1].

51. *Singularities of logarithmic kernels.* Integral equations with logarithmic kernels are also of interest in view of the applications we shall have to make of them in Chapter IV (Theory of Composition). Here, however, we shall limit our study to the equation

$$\int_0^x [\log(x-t) + C]y(t)\,dt = z(x) \qquad (1)$$

(C the Eulerian constant $0\cdot57721\ldots$); for a full discussion the reader is referred to VOLTERRA's works [2].

---

[1] Cf. EVANS, (14)—(16). Cf. also PICARD, (45).
[2] Cf. VOLTERRA, (42) and (44) of Bibliography IV.

Equation (1) can also be written in the form

$$\int_0^x y(t) \left[\frac{\partial}{\partial \alpha} \frac{(x-t)^{\alpha-1}}{\Gamma(\alpha)} \right]_{\alpha=1} dt = z(x), \qquad (2)$$

and, putting

$$z(x, \alpha) = \frac{\partial}{\partial \alpha} \psi(x, \alpha), \quad z(x, 1) = z(x),$$

we can write

$$\int_0^x y(t) \frac{(x-t)^{\alpha-1}}{\Gamma(\alpha)} dt = \psi(x, \alpha). \qquad (3)$$

Further, the function

$$\lambda(\xi, x) = \int_0^\infty \frac{(x-\xi)^\eta}{\Gamma(1+\eta)} d\eta$$

is finite and continuous; hence, multiplying both sides of (3) by $\lambda d\xi$ and integrating from 0 to x, we shall have

$$\int_0^x \psi(\xi, \alpha) \lambda(\xi, x) d\xi = \int_0^x \lambda(\xi, x) d\xi \int_0^\xi y(t) \frac{(\xi-t)^{\alpha-1}}{\Gamma(\alpha)} dt$$

$$= \int_0^x y(t) dt \int_t^x \lambda(\xi, x) \frac{(\xi-t)^{\alpha-1}}{\Gamma(\alpha)} d\xi$$

$$= \int_0^x y(t) dt \int_0^\infty d\eta \int_t^x \frac{(x-\xi)^\eta}{\Gamma(1+\eta)} \cdot \frac{(\xi-t)^{\alpha-1}}{\Gamma(\alpha)} d\xi$$

$$= \int_0^x y(t) dt \int_0^\infty \frac{(x-t)^{\alpha+\eta}}{\Gamma(1+\alpha+\eta)} d\eta,$$

and differentiating with respect to α,

$$\int_0^x z(\xi, \alpha) \lambda(\xi, x) d\xi = \int_0^x y(t) dt \frac{\partial}{\partial \alpha} \int_0^\infty \frac{(x-t)^{\alpha+\eta}}{\Gamma(1+\alpha+\eta)} d\eta. \qquad (4)$$

But the second integral on the right is merely $\int_\alpha^\infty \frac{(x-t)^\eta}{\Gamma(1+\eta)} d\eta$, and therefore its derivative with respect to α is $-\frac{(x-t)^\alpha}{\Gamma(1+\alpha)}$;

hence (4) becomes

$$\int_0^x y(t) \frac{(x-t)^\alpha}{\Gamma(1+\alpha)} dt = -\int_0^x z(\xi, \alpha) \lambda(\xi, x) d\xi,$$

and for $\alpha = 1$,

$$\int_0^x y(t)(x-t) dt = -\int_0^x z(\xi) \lambda(\xi, x) d\xi,$$

from which we get finally, by differentiating twice,

$$y(x) = -\frac{d^2}{dx^2} \int_0^x z(\xi) d\xi \int_0^\infty \frac{(x-\xi)^\eta}{\Gamma(1+\eta)} d\eta,$$

a formula which solves equation (1). In this case too we have applied the method of *transformation of the kernel* to solve this equation with a singular kernel.

Other interesting cases of particular integral equations can be found in VOLTERRA's works ([1]).

§ 9.

52. *Case in which the kernel vanishes for* $t = x$. If $K(x, x) = 0$ identically for any value of x whatever, the equations (2) und (4) of No. 48 are not equations of the second kind, but of the first; from the given equation

$$z(x) = \int_0^x K(x, t) y(t) dt \qquad (1)$$

we thus get, on differentiating once,

$$z'(x) = \int_0^x \frac{\partial K}{\partial x} y(t) dt, \qquad (2)$$

which is also an equation of the first kind. In order that (1) may have a solution, we must have not only $z(0) = 0$, but also, for (2), $z'(0) = 0$; and if the known functions are also differentiable, we shall have on differentiating again

$$z''(x) = \left(\frac{\partial K}{\partial x}\right)_{t=x} y(x) + \int_0^x \frac{\partial^2 K}{\partial x^2} y(t) dt; \qquad (3)$$

([1]) Cf. VOLTERRA, (42) and (44) of Bibliography IV.

and if in this equation $\left(\dfrac{\partial K}{\partial x}\right)_{t=x} \neq 0$ always, we shall again have an equation of the second kind, while if it is identically 0 we shall continue to apply the same method until (if it is possible) we find a derivative which either does not vanish in the interval considered, or vanishes at isolated points.

In the last of these cases the problem can be reduced to studying an equation of the first kind of the type (1), in which, however, the function $K(x, x)$ vanishes at isolated points $x_1 < x_2 < \cdots$. If $x_1 > 0$, then for x variable in the interval $(0, x_1)$ the case can be reduced to an equation (2) of No. 48; while for $x > x_1$ it can be reduced to the case of an equation (1) for which $K(x, x)$ vanishes at the lower limit of the integral, $K(0, 0) = 0$; this last case is therefore the only one that is essentially new.

In this case ([1]), if we suppose, first, that $K(x, t)$ is a polynomial of degree n in x, and is also such that $K(x, x)$ has an n-fold root for $x = 0$, it can be shown that a necessary condition for the solution of (1) to exist and to be finite at the origin is that $z(x)$ shall have an $(n+1)$-fold root for $x = 0$. Differentiating equation (1) $n + 1$ times we shall then reach an *ordinary linear differential equation of order n*

$$a_0(x) y^{(n)}(x) + a_1(x) y^{(n-1)}(x) + \cdots$$
$$+ a_n(x) y(x) = z^{(n+1)}(x) \qquad (4)$$

which $y(x)$ must satisfy; and imposing on $y(x)$ the condition of being regular at the origin, we obtain, under certain conditions, one and only one solution of the differential equation which can be shown to be the unique solution of the integral equation (1). In the general case, when $K(x, t)$ is not a polynomial in x, but is still such that $K(x, x)$ has an n-fold root for $x = 0$, then by a similar procedure (i.e. differentiating (1) $n + 1$ times) we should reach, not a differential equation of order n in $y(x)$, but an equation of the *integro-differential* type

$$a_0(x) y^{(n)}(x) + a_1(x) y^{(n-1)}(x) + \cdots + a_n(x) y(x)$$
$$+ \int_0^x K_n(x, t) y(t) dt = z^{(n+1)}(x). \qquad (5)$$

([1]) Cf. LALESCO, (30) and (31); VOLTERRA, (76), Notes 3 and 4.

To determine y we should have to apply the method of successive approximations, determining first a $y_0(x)$ as before from the differential equation obtained from (5) by neglecting the integral, then a $y_1(x)$ from (5) by substituting y_0 for $y(t)$ under the integral sign, and so on. It can then be shown that the sequence of terms $y_i(x)$ so determined, as i increases to ∞, is uniformly convergent, in a certain interval, towards a function $y(x)$; we therefore have that, under certain hypotheses as to the differential equations to be integrated, the solution $y(x)$ of the equation (1) exists and is unique.

The above considerations, especially the part of the argument leading to the differential equation (4), bring out clearly, in a particular way, the close connection between ordinary *linear differential equations* and *Volterra's integral equations*.

§ 10. *Systems of Integral Equations; Inversion of Multiple Integrals.*

53. Systems of integral equations with either variable or fixed limits can be discussed in a way that is completely analogous to what has been seen above for equations with a single unknown function. Thus, for example, for the systems with variable limits of the type

$$z_i(x) = y_i(x) + \lambda \int_0^x \sum_1^n K_{ir}(x,t) y_r(t) dt \quad (i = 1, 2, \ldots n) \quad (1)$$

(a system of the second kind), we can prove that a system of solutions $y_i(x)$ exists and is unique; while for systems with constant limits we can prove that a system of solutions exists and is unique when λ is not a root of a certain integral transcendental $D(\lambda)$ (the determinant of the system) [1].

Further, in accordance with an observation of FREDHOLM, these systems of integral equations are always reducible to a single integral equation with a discontinuous kernel.

The same methods also apply to the case — of particular interest in view of its applications to mathematical physics — of linear integral equations in which the unknown is a function

---

[1] Cf. FREDHOLM, (20).

II THE FUNCTIONAL CALCULUS 63

of several variables. Thus, for instance, for the integral equation

$$z(x_1, x_2) = y(x_1, x_2) + \lambda \int_a^b \int_a^b K(x_1, x_2; t_1, t_2) y(t_1, t_2) dt_1 dt_2, \quad (2)$$

or for the other equivalent equation

$$z(x_1, x_2) = y(x_1, x_2) + \lambda \int_\sigma K(x_1, x_2; t_1, t_2) y(t_1, t_2) dt_1 dt_2$$

(in which the area of integration σ is not a square, but which can be brought within the previous case by circumscribing a square to σ and supposing the kernel zero at all points of the square outside σ), we can construct two integral functions in λ, $D_k(\lambda)$, the determinant of the equation, and $D_k(\lambda; x_1, x_2; t_1, t_2)$. When λ is not a root of $D_k(\lambda)$ (i.e. is not a *characteristic value*), the solution, which is unique, of the equation is expressed by means of a double integral of the quotient $\dfrac{D_k(\lambda; x_1, x_2; t_1, t_2)}{D_k(\lambda)}$. If, on the other hand, $D_k(\lambda) = 0$, the solution of (2) does not always exist, and when it exists it is not unique; but some solution certainly exists for the *homogeneous* equation obtained from (2) by putting $z(x_1, x_2) = 0$.

The case of triple integrals, etc., can be treated in a similar way.

§ 11. *General Functional Equations and Implicit Functionals.*

54. Having examined some of the principal linear functional equations, we can now turn to the general case

$$\mathrm{F}[y(t); x] = z(x); \quad (1)$$

and if we suppose that this equation is satisfied for a pair of functions $y_0(t), z_0(x)$, we can try to determine what variation $\delta y(t)$ must be added to $y_0(t)$ in order that (1) may still be satisfied when we make $z_0(x)$ undergo a variation $\delta z(x)$. To determine the variation $\delta y(t)$ we shall have the other functional equation

$$\delta \mathrm{F}[y(t); x]_{y=y_0} = \delta z(x), \quad (2)$$

which is a linear equation (see above, No. 42) with respect to the unknown $\delta y(t)$. For the equations (1) we can therefore adopt a classification based on the type of the corresponding linear equation (2) (the equation of variations).

Thus, for instance, if equation (1) is of the type

$$\mu z(x) = y(x) + \lambda \int_0^1 K_1(x,t) y(t) \, dt + \cdots$$
$$+ \frac{\lambda^n}{n!} \int_0^1 \int_0^1 \cdots \int_0^1 K_n(x, t_1, t_2, \ldots t_n) y(t_1) y(t_2) \ldots$$
$$\ldots y(t_n) \, dt_1 \, dt_2 \ldots dt_n + \cdots \tag{3}$$

(with the kernels K_n symmetrical with respect to the t_i's), the corresponding equation of variations will be, for $z_0(x) = y_0(t) = 0$,

$$\mu \delta z(x) = \delta y(x) + \lambda \int_0^1 K_1(x,t) \delta y(t) \, dt, \tag{4}$$

an *integral* equation *of the second kind with constant limits*; and we shall apply the same denomination to the transcendental equation (3) from which it is derived.

In order to solve (3) ([1]), we shall try to determine y as a function of μ, in such a way that it satisfies (3) itself, and that $y = 0$ for $\mu = 0$. Differentiating (3) once with respect to μ and putting $\mu = 0$, we shall have

$$z(x) = \left[\frac{dy}{d\mu}\right]_{\mu=0} + \lambda \int_0^1 K_1(x,t) \left[\frac{dy}{d\mu}\right]_{\mu=0} dt, \tag{5}$$

a linear integral equation of the second kind in the unknown function $\left[\dfrac{dy(t)}{d\mu}\right]_{\mu=0}$. If we suppose the determinant $D_{k_1}(\lambda)$ of the equation different from 0, we can then find one and only one function $\left[\dfrac{dy(t)}{d\mu}\right]_{\mu=0}$. Differentiating (3) twice with respect to μ and then putting $\mu = 0$, we shall have

$$0 = \left[\frac{d^2 y}{d\mu^2}\right]_{\mu=0} + \lambda \int_0^1 K_1(x,t) \left[\frac{d^2 y}{d\mu^2}\right]_{\mu=0} dt$$
$$+ \lambda^2 \int_0^1 \int_0^1 K_2(x, t_1, t_2) \left[\frac{dy(t_1)}{d\mu}\right]_{\mu=0} \left[\frac{dy(t_2)}{d\mu}\right]_{\mu=0} dt_1 \, dt_2,$$

an integral equation of the second kind in the unknown function $\left[\dfrac{d^2 y}{d\mu^2}\right]_{\mu=0}$, from which, having supposed $D_{k_1}(\lambda) \neq 0$, we

---

([1]) Cf. VOLTERRA, (80); cf. also (84) of Bibliography I.

can obtain $\left[\dfrac{d^2 y}{d\mu^2}\right]_{\mu=0}$, and so on. Having then constructed the series

$$y(x) = \mu \left[\frac{dy}{d\mu}\right]_{\mu=0} + \frac{\mu^2}{2!}\left[\frac{d^2 y}{d\mu^2}\right]_{\mu=0} + \cdots + \frac{\mu^n}{n!}\left[\frac{d^n y}{d\mu^n}\right]_{\mu=0} + \cdots, \quad (6)$$

we can show that it is uniformly convergent if $|\mu z(x)| < \varepsilon$, with ε conveniently small, and in this region round $\mu z_0 = 0$ it represents the only solution $y(x)$ of (3) that can be obtained with continuity from the solution $y_0 = 0$ corresponding to $\mu z_0 = 0$. This theorem corresponds exactly, in the functional calculus, to the theorem of the ordinary calculus on the existence of implicit functions; and as, in the ordinary calculus, for a system

$$f_i(y_1, y_2, \ldots y_n) = z_i \quad (i = 1, 2, \ldots n)$$

we exclude the possibility that the functional determinant $\dfrac{\partial(f_1, f_2, \ldots f_n)}{\partial(y_1, y_2, \ldots y_n)}$, corresponding to the initial solution, may vanish, so here, in the above considerations, we have had to exclude the possible vanishing of the determinant $D_{k_1}(\lambda)$ of the linear equation (5); so that this determinant has in the functional calculus the same importance and the same properties as the functional determinant in the ordinary calculus.

55. More precisely, SCHMIDT[1] has shown that if corresponding to a solution $y_0(t)$ of the functional equation

$$F\left[\underset{a}{\overset{b}{y(t)}};x\right] = z(x), \quad F[y_0(t);x] = z_0(x) \qquad (7)$$

we have an equation of variations of the type

$$\delta z(x) = \delta y(x) + \lambda_0 \int_0^b K(x,t)\,\delta y(t)\,dt,$$

and its determinant $D_k(\lambda_0) = 0$, then the function $z_0(x)$ is, with certain complementary hypotheses, a *branch element* for the functional

$$y(x) = G[z(x);t]$$

---

[1] Cf. SCHMIDT, (60).

defined implicitly by the equation (1), and *vice versa*; or, in other words, that if $z(x)$ is taken within a suitably restricted region round $z_0(x)$, there exist *several functions* $y(t)$ of the region round $y_0(t)$ which when substituted in (7) satisfy it.

56. If in the equation (7) the functional F depends on the values taken by y when t varies *only in the interval* $(0, x)$, the corresponding equation of variations will be a *Volterra's integral equation*, and therefore its determinant D_k will always be different from zero. We may therefore logically expect to find, for the implicit function $y(t) = \mathrm{G}[z(x); t]$ defined by

$$\mathrm{F}\left[\underset{0}{\overset{x}{y(t)}}; x\right] = z(x), \tag{8}$$

that it is impossible to have two or more branch elements $z(x)$; and in fact SABATINI ([1]) has shown that with this hypothesis the solution $y(t)$ of (8) is *always unique*.

We shall briefly summarise the proof for the case of a functional F of the second degree, i.e. of an equation

$$y(x) + \int_0^x K_1(x, t) y(t) dt$$
$$+ \int_0^x \int_0^x K_2(x, t_1, t_2) y(t_1) y(t_2) dt_1 dt_2 = z(x).$$

If there could exist two solutions $y_1(x)$, $y_2(x)$ of this equation, then putting $u(x) = y_1(x) - y_2(x)$ we should have

$$u(x) + \int_0^x K_1(x, t) u(t) dt$$
$$+ \int_0^x \int_0^x K_2(x, t_1, t_2) y_1(t_1) u(t_2) dt_1 dt_2$$
$$+ \int_0^x \int_0^x K_2(x, t_1, t_2) y_2(t_2) u(t_1) dt_1 dt_2 = 0,$$

i.e. supposing K_2 symmetrical in t_1 and t_2, and putting

$$H(x, t) = K_1(x, t) + \int_0^x K_2(x, t_1, t) [y_1(t_1) + y_2(t_1)] dt_1,$$

([1]) Cf. SABATINI, (54).

we should have

$$u(x) + \int_0^x H(x,t)u(t)\,dt = 0,$$

and this *homogeneous* VOLTERRA's equation has no solution other than $u(x) = 0$; hence $y_1(x) = y_2(x)$.

§ 12. Calculation of Solvent Kernels and Approximate Solution.

57. Before leaving this part of the discussion of integral equations, we may indicate some methods of actually calculating the *solvent kernel* $S(x, t)$, or the associate of a kernel $K(x, t)$, of a *linear* equation of the second kind

$$z(x) = y(x) + \int_0^x K(x,t)y(t)\,dt.$$

By equation (11) of No. 43, we have

$$\int_x^t K(x,\xi)S(\xi,t)\,d\xi = K(x,t) + S(x,t). \tag{1}$$

If we suppose that $K(x, t)$ satisfies a homogeneous differential equation of the type

$$\sum_{i=0}^n a_i(x)\frac{\partial^i}{\partial x^i}K(x,t) = 0, \tag{2}$$

we shall have, differentiating (1) i times with respect to x,

$$\int_x^t \frac{\partial^i K(x,\xi)}{\partial x^i}S(\xi,t)\,d\xi - \sum_{s=0}^{i-1}\frac{\partial^s}{\partial x^s}\left\{\left[\frac{\partial^{i-1-s}}{\partial x^{i-1-s}}K(x,t)\right]_{t=x} S(x,t)\right\}$$

$$= \frac{\partial^i}{\partial x^i}K(x,t) + \frac{\partial^i}{\partial x^i}S(x,t) \qquad (i = 0, 1, 2, \ldots n). \tag{3}$$

Multiplying these equations by $a_i(x)$ and summing we get a new equation in which the term with the integral vanishes because of (2), and there remains a *linear differential equation of order* n *in the solvent kernel* $S(x, t)$, from which we can determine $S(x, t)$ with the initial conditions (for $x = t$) given by the equations (3).

This method, which is due to EVANS [1], is particularly useful in the case of the so-called *closed cycle* (cf. No. 87), in which both $K(x, t)$ and its solvent kernel are functions of the difference $x - t$. Thus, for example, if $K(x, t) = -\sin(x - t)$, then

$$\frac{\partial^2 K}{\partial x^2} + K = 0,$$

and for the reciprocal kernel $S(x, t)$ it easily follows that

$$S(x, t) = \frac{\sqrt{2}}{2} \sin\left[\sqrt{2}\,(x - t)\right].$$

This case of the closed cycle has been further developed in a recent memoir by WHITTAKER [2].

TEDONE [3] has also discussed the problem of determining those kernels $K(x, t)$ (also for equations with variable limits) whose reciprocal kernels can be calculated by means of elementary operations and operations of differentiation and integration.

Lastly, various methods have been proposed for the approximate numerical solution of linear integral equations.

---

[1] Cf. EVANS, (17).
[2] Cf. WHITTAKER, (84).
[3] Cf. TEDONE, (63); VOLTERRA, (83).

BIBLIOGRAPHY II(¹).

(1) ABEL (N. H.).—Résolution d'un problème de mécanique. (*Œuvres*, Vol. I, p. 97.)
(2) ABEL (N. H.).—Solutions de quelques problèmes à l'aide d'intégrales définies. (*Œuvres*, Vol. I, p. 113.)
(3) BENNETT (A. C.).—Newton's method in general analysis. (*Proc. Nat. Ac. of Sc.*, Vol. II, 1916.)
(4) BERNSTEIN (F.).—Die Integralgleichung der elliptischen Thetanullfunktion. (*Berliner Berichte*, 1920.)
(5) BLOCK (H.).—Sur la solution de certaines équations fonctionnelles. (*Ark. Math.* Stockholm, Vol. 3, No. 22, 1907.)
(6) BÔCHER (M.).—An Introduction to the Study of Integral Equations. Cambridge, 1909; 2nd ed., Cambridge, 1914.
(7) BROWNE (P. J.).—Integral Equation proposed by Abel. (*Proc. Royal Irish Academy*, Vol. XXXII, Sec. 4, No. 6; Dublin, 1915.)
(8) BROWNE (P. J.).—Sur quelques cas singuliers de l'équation de Volterra. (*C. R. de l'Acad. des Sciences de Paris*, Vol. 154, 1912.)
(9) BUCHT (G.).—Über nichtlineare Integralgleichungen mit unverzweigten Lösungen. (*Ark. Math. Stockholm*, Bd. 8, 1912.)
(10) BURGATTI (M.).—Sull'inversione degli integrali definite. (*R. Acc. dei Lincei. Rend.*, 1903 (2 notes).
(11) DAVIS (H. T.).—The present status of integral equations. (*Indiana University Studies*, No. 70, Vol. XIII, 1926.)
(12) DAVIS (H. T.).—A survey of methods for the inversion of integrals of Volterra type. (*Indiana University Studies*, Nos. 76, 77, 1927.)
(13) DIXON (A. C.).—Note on functional equations which are limiting forms of integral equations. (*London Math. Soc. Proc.*, Vol. 17, 1919.)
(14) EVANS (G. C.).—The integral equation of the second kind of Volterra, with singular kernel. (*Amer. Math. Soc. Bull.*, 2nd series, Vol. XVI, 1909.)
(15) EVANS (G. C.).—Volterra's integral equation of the second kind with discontinuous kernel. (*Amer. Math. Soc. Trans.*, Vol. XI, 1910; Vol. XII, 1911.)

(¹) We give only a concise bibliography on integral equations. A more complete bibliography is to be found in the works mentioned here; cf., for instance, (11).

(16) EVANS (G. C.).—L'equazione integrale di Volterra di seconda specie con un limite dell'integrale infinito. (*R. Acc. dei Lincei. Rend.*, Vol. XX, 1911.)

(17) EVANS (G. C.).—Sul calcolo del nucleo dell'equazione risolvente per una data equazione integrale. (*R. Acc. dei Lincei. Rend.*, Vol. XX, 1911.)

(18) EVANS (G. C.).—Some general types of functional equations. (*Inter. Congress of Math.*, Cambridge, 1912.)

(19) FANTAPPIÈ (L.).—Risoluzione di una classe di equazioni integrali di la specie a limiti costanti. (*R. Acc. dei Lincei. Rend.*, Series 6, Vol. I, 1925.)

(20) FREDHOLM (I.).—Sur une classe d'équations fonctionnelles. (*Acta Math.*, Vol. 27, 1903.)

(21) GOURSAT (E.).—Cours d'analyse mathématique (Vol. III: Équations intégrales). 3rd ed. Gauthier-Villars, Paris, 1923.

(22) HADAMARD (J.).—La série de Taylor et son prolongement analytique. (Coll. Scient., No. 12, Paris, 1901.)

(23) HILBERT (D.).—Grundzüge einer allgemeinen Theorie der linearen Integralgleichungen. (Teubner, Leipzig, 1912.)

(24) HOLMGREN (E.).—Sur un théorème de M. Volterra. (*Atti Torino.*, Vol. XXXV, 1900.)

(25) HOLMGREN (H.).—Om Differentialkalkylen med indices af hvad Natur som helst. (*Stockholm, Ak. Handl.*, Vol. V, 1863-64.)

(26) HORN (I.).—Über nichtlineare Integralgleichungen vom Volterra-'schen Typus. (*Deutsche Math. Ver.*, Bd. 27, 1918.)

(27) HORN (I).—Singuläre Systeme linearer Volterra'schen Integralgleichungen. (*Math. Zeitschr.*, Vol. 3, 1919.)

(28) INCE (E. P.).—On the connexion between linear differential systems and integral equations. (*Proc. R. Soc. Edinburgh*, Vol. 42.)

(29) KAKEYA (S.).—On an infinite number of linear integral equations. (*Sc. Rep. Tohoku Imp. Uni. Sendai*, Vol. 5, 1916.)

(30) LALESCO (T.).—Sur l'équation de Volterra. (Thèse.) (Gauthier-Villars, Paris, 1908.)

(31) LALESCO (T.).—Sur l'équation de Volterra. (*Journal de Math.*, Vol. I, 1908.)

(32) LALESCO (T.).—Sur une équation du type Volterra. (*C. R. de l'Acad. de Scs. de Paris*, 1911.)

(33) LALESCO (T.).—Introduction à la théorie des équations intégrales. (Paris, 1912.)

(34) LAURICELLA (G.).—Sopra alcune equazioni integrali. (*R. Acc. dei Lincei. Rend.*, Series 5, Vol. XVII, 1908; Vol. XVIII, 1908 (3 notes).)

(35) LAURICELLA (G.).—Sulla resoluzione dell'equazione integrale di la specie. (*R. Acc. dei Lincei. Rend.*, Series 5, Vol. XX, 1911.)

(36) LE ROUX (F.).—Sur les intégrales des équations linéaires aux dérivées partielles du second ordre à deux variables indépendants. (*Thèse.* Paris, 1894.—*Ann. Ec. Norm.*, 1895.)

(37) LEVI-CIVITA (T.).—Sull'inversione degli integrali definiti nel campo reale. (*Atti Acc. di Torino*, Vol. XXXI, 1895.)

(38) LÉVY (P.).—Sur les fonctionnelles implicites. (*Bull. de la Soc. Math. de France*, 1919.)

(39) LÉVY (P.).—Sur les fonctions de lignes implicites. (*C. R. de l'Ac. des Sciences de Paris*, Vol. 168, 1919. *Bull. de la Soc. Math. de France*, Vol. 48, 1920.)
(40) LIOUVILLE (J.).—Mémoire sur quelques questions de géométrie et de mécanique et sur un nouveau genre de calcul pour résoudre ces questions. (*J. de l'École Polytechnique*, XXIième Cahier, 1832.)
(41) LIOUVILLE (J.).—Mémoire sur le calcul des différentielles à indices quelconques. (*J. de l'École Polytechnique*, XXIième Cahier, 1832.)
(42) LIOUVILLE (J.).—Mémoire sur l'intégration de l'équation
$(m x^2 + n x + p) \dfrac{d^2 y}{d x^2} + (q x + r) \dfrac{d y}{d x} + s y = 0$, à l'aide des différentielles à indices quelconques. (*J. de l'École Polytechnique*, XXIième Cahier, 1832.)
(43) MANDELBROJT (S.).—Sulla generalizzazione del calcolo delle variazioni. (*R. Acc. dei Lincei. Rend*, Series 6, Vol. 1, 1925.)
(44) MOLINARI (A. M.).—Derivazione a indice qualunque. (*R. Acc. dei Lincei. Rend.*, Series 5, Vol. 25-26, 1916-1917.)
(45) PICARD (E.).—Sur une équation fonctionnelle. (*C. R. de l'Acad. des Sciences de Paris*, 1904, 2ième sem.)
(46) PICARD (E.).—Sur une équation fonctionnelle. (*C. R. de l'Acad. des Sciences de Paris*, 1907, 1er sem.)
(47) PICARD (E.).—Quelques remarques sur les équations intégrales de première espèce et sur certains problèmes de physique mathématique. (*C. R. de l'Acad. des Sciences de Paris*, Vol. 148, 1909.)
(48) PICARD (E.).—Sur les équations intégrales de première espèce. (*C. R. de l'Acad. des Sciences de Paris*, Vol. 148, 1909.)
(49) PICARD (E.).—Sur les équations intégrales de troisième espèce. (*Ann. de l'École Normale*, 3ième ser., Vol. 28. Paris, 1911.)
(50) PICONE (M.).—Sopra un'equazione integrale di prima specie a limiti variabili considerata da Volterra. (*R. Acc. dei Lincei. Rend.*, Series 5, Vol. XIX, 1910.)
(51) POLI (C.).—Un theorema di esistenza per equazioni integrali non lineari. (*Atti Acc. di Torino*, Vol. 51, 1915-16.)
(52) PRASAD (GORAKH).—On the numerical solution of integral equations. (*Abstr. Intern. Congress of Toronto*, 1924.)
(53) RIEMANN (B.).—Versuch einer allg. Auffassung der Integration und Differentiation. 1847. (Werke, Leipzig, 1892.)
(54) SABBATINI (A.).—Sull'unicità della soluzione delle equazioni integrali a limiti variabili. (*R. Accad. dei Lincei. Rend.*, Series 6, Vol. 9, 1925.)
(55) SABBATINI (A.).—Sulle equazioni integrali quadratiche a limiti variabili..(*R. Acc. dei Lincei. Rend.*, Series 6, Vol. 1, 1925.)
(56) SBRANA (F.).—Sopra certe equazioni integrali considerate dal prof. Tedone. (*R. Acc. Lincei. Rend.*, Series 5, Vol. XXX, 1921.)
(57) SBRANA (F.).—Sopra alcune formule di risoluzione di certe equazcioni integrali di Volterra. (*R. Acc. dei Lincei. Rend.*, Series 5, Vol. XXXI, 1922.)
(58) SBRANA (F.).—Sopra certe equazioni integrali di Volterra risolubili con procedimenti finiti. (*R. Acc. dei Lincei. Rend.*, Series 5, Vol. XXXII, 1923.)

(59) SCATIZZI (S.).—Soluzione di qualche tipo di equazioni differenziali ad indice qualunque. (*R. Acc. dei Lincei. Rend.*, Series 5, Vol. XXXI, 1922; Series 5, Vol. XXXII, 1923.)

(60) SCHMIDT (E.).—Zur Theorie der linearen und nichtlinearen Integralgleichungen. (*Math. Ann.*, Bd. 63, 64, 65, 1906-07-08.)

(61) SONINE (N.).—Recherches sur les fonctions cylindriques. (*Math. Ann.*, Bd. XVI.)

(62) SONINE (N.).—Sur la généralisation d'une formule d'Abel. (*Math. Ann.*, Bd. IV.)

(63) TEDONE (O.).—Su alcune equazioni integrali di Volterra risolubili con un numero finito di derivazioni e di integrazioni. (*R. Acc. dei Lincei. Rend.*, Series 5, Vol. XXIII, 1914.)

(64) USAI (G.).—Sulle soluzioni in termini finiti di equazioni integrali col nucleo $x-y$. (*Cir. Mat. di Palermo, Rend.*, Vol. 45, 1921.)

(65) VERGERIO (A.).—Sull'equazioni integrali di Fredholm di 1a specie. (*Ist. Lombardo. Rend.*, Vol. 47, 1914.)

(66) VERGERIO (A.).—Sulla risolubilità dell'equazione integrale di la specie. (*R. Acc. de Lincei. Rend.*, Series 5, Vol. XXIV, 1915.)

(67) VERGERIO (A.).—Una condizione necessaria e sufficiente per l'esistenza di soluzioni nell'equazione integrale di prima specie. (*R. Acc. dei Lincei. Rend.*, Series 5, Vol. XXIV, 1915.)

(68) VERGERIO (A.).—Sulle equazioni integrali non lineari. (*Cir. Mat. di Palermo. Rend.*, Vol. XLI, 1916.)

(69) VERGERIO (A.).—Sulle equazioni integrali non lineari con operazioni funzionali singolari. (*Giorn. di Mat. di Battaglini*, Vol. 59, 1921.)

(70) VERGERIO (A.).—Sulle equazioni integrali non lineari. (*Ann. di Mat. di Brioschi*, Series 3, Vol. 31, 1922.)

(71) VERGERIO (A.).—Sopra un tipo di equazioni integrali non lineari. (*R. Acc. dei Lincei. Rend.*, Series 5, Vol. XXXI, 1922.)

(72) VITERBI (A.).—Su la risoluzione approssimata delle equazioni integrali di Volterra e su la applicazione di queste a lo studio analitico delle curve. (*Ist. Lombardo. Rend.*, Series 2, Vol. 45, 1912.)

(73) VIVANTI (G.).—Elementi della teoria delle equazioni integrali lineari. (Manuali Hoepli. Milano, 1916.)

(74) VOLTERRA (V.).—Sopra un problema di elettrostatica. (*Nuovo Cimento*, Series 3, Vol. XVI, 1884.)

(75) VOLTERRA (V.).—Sopra un problema di elettrostatica. (*R. Acc. dei Lincei. Trans.*, 1, 3°, Vol. VIII, 1884.)

(76) VOLTERRA (V.).—Sulla inversioni degli integrali definiti. (*Atti R. Acc. Torino*, Vol. XXXI, 1896.) (Equazioni integrali di 1a specie a limiti variabili.) Notes 1, 2, 3, and 4. (Ibid.)

(77) VOLTERRA (V.). — Sulla inversione degli integrali definiti. (*R. Acc. dei Lincei. Rend.*, Series 5, Vol. V, 1896.)

(78) VOLTERRA (V.).—Sulla inversione degli integrali multipli. (*R. Acc. dei Lincei. Rend.*, Series 5, Vol. V, 1896.)

(79) VOLTERRA (V.).—Sopra alcune questioni d'inversione di integrali definiti. (*Ann. di Mat.* Milano, 1897.)

(80) VOLTERRA (V.).—Sur les fonctions qui dépendent d'autres fonctions. (*C. R. de l'Acad. des Sciences de Paris*, 19 May 1906.)

(81) VOLTERRA (V.).—Sopra equazioni di tipo integrale. (*Intern. Congress of Math.*, Cambridge, 1912.)
(82) VOLTERRA (V.).—Leçons sur les équations intégrales et les équations intégro-différentielles. (Coll. Borel.) (Gauthier-Villars, Paris, 1913.)
(83) VOLTERRA (V.).—Osservazioni sui nuclei delle equazioni integrali. (*R. Acc. dei Lincei. Rend.*, Series 5, Vol. XXIII, 1914.)
(84) WHITTAKER (E. C.).—On the numerical solution of integral equations. (*Proc. Royal Soc.*, Vol. 94, Series A, 1918.)
(85) ZEILON (Nils).—Sur quelques points de la théorie de l'équation intégrale d'Abel. (*Arkiv för Mat. Astro. och Fys. av. k. Svenska Vetenskapsakad.*, Vol. 18, Stockholm, 1924.)

CHAPTER III

GENERALISATION OF ANALYTIC FUNCTIONS.

§ 1.

58. Starting from the ordinary theory of analytic functions, we know that there are three principal methods of developing it, namely:

(i) the method of LAGRANGE-WEIERSTRASS (expansions in power series);

(ii) CAUCHY's method (conditions of monogeneity);

(iii) RIEMANN's method (conjugate harmonic functions; differential parameters).

These three methods in general are not developed separately, but they supplement one another. They are all capable of leading to various extensions of the theory of analytic functions, such, for instance, as power series and analytic functions of several variables[1]; but the extension of the concept of an analytic function which we now propose to consider is closely connected with the *theory of functionals*, or rather, more precisely, with *functions of lines* [2] (or more generally of *hyperspaces*), as defined above (No. 9).

---

[1] See also the extension of the theory of functions developed in the thesis of Dr. M. NICOLESCO, (8).

[2] The name *function of a line* was initially taken to mean what in general is now called a *functional*. In this sense the term "function of a line" has been used by many writers, and in particular by VOLTERRA, who was the first to introduce this concept, in his Paris lectures (cf. bibliography to Chapter I, (87)) and in many earlier works. At present, however, having adopted the term "functional" for this general concept as being more convenient because less specific, we shall reserve the name "functions of lines" for those particular functions, of a more strictly geometrical nature, which we are now about to define. Cf. also No. 9 of the text.

CHAP. III   ANALYTIC FUNCTIONS   75

59. Starting from the definition just referred to, we shall say that a quantity is a *function of a line* (closed and not intersecting) in space when to each of such lines (which we shall always suppose *rectifiable*) L there can be made to correspond a definite value of the quantity, which we shall denote by $F[L]$. These *functions of lines* $F[L]$ will evidently be particular *functionals* of the three functions $x(s)$, $y(s)$, $z(s)$, which give the coordinates x, y, z of the points of the line L as s varies between certain limits 0 and S, where s denotes the length of the arc between a variable point of L and a fixed point of L taken as origin; i.e.

$$F[L] = F\left[\underset{0}{\overset{S}{x(s)}}, \underset{0}{\overset{S}{y(s)}}, \underset{0}{\overset{S}{z(s)}}\right],$$

where S denotes the length of the curve L.

If we suppose the two functions $y(s)$ and $z(s)$ fixed, and if we give a variation δx to the function $x(s)$ (which is equivalent to displacing the curve L along a cylinder whose generators are parallel to the axis of x), the functional

$$F[L] = F\left[\underset{0}{\overset{S}{x(s)}}, \underset{0}{\overset{S}{y(s)}}, \underset{0}{\overset{S}{z(s)}}\right],$$

which we shall suppose regular, will undergo a variation $\delta_x F$ expressed by

$$\delta_x F = \int_0^S F'_x[L;s]\delta x\,ds,$$

where we have put $F'_x[L;s]$ to denote the functional derivative of the functional $F[L] = F\left[\underset{0}{\overset{S}{x(s)}}, \underset{0}{\overset{S}{y(s)}}, \underset{0}{\overset{S}{z(s)}}\right]$ with respect to the function $x(s)$ at the point s (Nos. 25 to 28).

Similarly we shall have, if y alone varies by δy, and z alone by δz,

$$\delta_y F = \int_0^S F'_y[L;s]\delta y\,ds,$$

$$\delta_z F = \int_0^S F'_z[L;s]\delta z\,ds;$$

and if we denote simply by X, Y, Z, the three functional derivatives F'_x, F'_y, F'_z of $F[L]$ (which are also, in general, functions of the line L, and functions of the point S), the total variation δF of $F[L]$ will be given by (Nos. 26 *et seq.*)

$$\delta F = \int_0^S (X\delta x + Y\delta y + Z\delta z)\,ds. \tag{1}$$

60. We have however already said that these functions of lines $F[L]$ are *particular* functionals of the three functions $x(s)$, $y(s)$, $z(s)$ which determine L; in fact, if we move the points of the line L along L itself, the functional F must not change, since it depends only on the line L and not on the position of the separate points along L. This can also be expressed by saying that if to $x(s)$, $y(s)$, $z(s)$ we give variations $\delta x = k(s)\alpha(s)$, $\delta y = k(s)\beta(s)$, $\delta z = k(s)\gamma(s)$ ($k(s)$ infinitesimal) proportional to the direction cosines α, β, γ of the tangent to L at the point s (which is equivalent to a displacement of L along itself), the functional $F[L]$ must not change its value, so that we shall have

$$\delta F = \int_0^S k(s)(X\alpha + Y\beta + Z\gamma)\,ds = 0, \tag{2}$$

and since $k(s)$ is arbitrary, the three functional derivatives X, Y, Z of $F[L]$ at the point S with respect to $x(s)$, $y(s)$, $z(s)$ must satisfy the relation

$$X\alpha + Y\beta + Z\gamma = 0. \tag{3}$$

To satisfy this relation, we need only put

$$X = B\gamma - C\beta, \quad Y = C\alpha - A\gamma, \quad Z = A\beta - B\alpha; \tag{4}$$

i.e. we consider the three functional derivatives X, Y, Z as the three components of the vector product of the two vectors (α, β, γ) and (A, B, C), the second of which is not completely determined, since we can replace it by another of components A', B', C' given by

$$A' = A + K\alpha, \quad B' = B + K\beta, \quad C' = C + K\gamma.$$

With the transformation (4), the expression (1) for δF takes the form

$$\delta F = \int_0^S [A(\beta\delta z - \gamma\delta y) + B(\gamma\delta x - \alpha\delta z) + C(\alpha\delta y - \beta\delta x)]\,ds,$$

or, denoting by ds_x, ds_y, ds_z, respectively the three components αds, βds, γds, of the vector ds (of length ds and direction the same as that of the tangent), we shall have

$$\delta F = \int [A(ds_y \delta z - ds_z \delta y) + B(ds_z \delta x - ds_x \delta z) + C(ds_x \delta y - ds_y \delta x)];$$

and if we denote by $d\sigma$ the area of the infinitesimal parallelogram described by the vector ds when displaced by an amount $\delta l = (\delta x, \delta y, \delta z)$ from its initial position, and by $\cos nx$, $\cos ny$, $\cos nz$ the direction cosines of the normal to this parallelogram, then the expressions in round brackets in the preceding expression will be precisely the components $d\sigma \cos nx$, $d\sigma \cos ny$, $d\sigma \cos nz$, of the vectorial product $ds \wedge \delta l$, and we shall therefore have also

$$\delta F = \int (A \cos nx + B \cos ny + C \cos nz) d\sigma, \tag{5}$$

where the integral is taken along the infinitesimal ring-shaped surface described by the line L when its points are displaced by the amount δl.

If now we make the closed curve L vary continuously from an initial position L_0 to another L_1, it will describe a surface σ bounded by L_0 and L_1, and the difference $F[L_1] - F[L_0]$ of the function of the line will, by the preceding formula, be given by

$$F[L_1] - F[L_0] = \int_\sigma (A \cos nx + B \cos ny + C \cos nz) d\sigma;$$

and if, in particular, the initial curve L reduces to a point and corresponding to this the function of the line $F[L_0]$ vanishes, we shall have

$$F[L_1] = \int_\sigma (A \cos nx + B \cos ny + C \cos nz) d\sigma, \tag{6}$$

where σ is now a portion of the surface which has the curve L_1 for its complete boundary.

§ 2.

61. Now let us suppose that the quantities A, B, C, which are in general *functions of the curve L* and *of the point* on the

curve itself, and are not completely determined, can be chosen in such a way that they *do not depend on the curve L* (are *not* functionals of $x(s)$, $y(s)$, $z(s)$); that is to say, we are supposing that there exist three *functions* A, B, C of the point in space alone (functions of the coordinates x, y, z of the point) which can be used, as in the preceding section, whatever may be the closed curve L through the point. If we interpret A, B, and C as the three components of a vector, itself a function of the point in space, formula (6) of No. 60 will represent the value $F[L_1]$ of our function of the line as the *flux* of this vector across a surface σ which has the curve L_1 for its complete boundary; and since this quantity $F[L_1]$ must depend only on the curve L_1 and not on the surface σ, it follows that if σ' is another surface which also has L_1 for its complete boundary, we must have

$$\int_\sigma (A\cos nx + B\cos ny + C\cos nz)\,d\sigma$$
$$= \int_{\sigma'} (A\cos nx + B\cos ny + C\cos nz)\,d\sigma' = F[L_1]. \quad (1)$$

From this it follows that the flux of the vector (A, B, C) across any *closed* surface whatever (as, for example, the aggregate of the two surfaces σ and σ') must be zero. But a closed surface round any point whatever can be arbitrarily restricted, and therefore the *divergence* of the vector must also vanish at every point of the region considered, so that we must have identically, at every point of this region ([1]),

$$\frac{\partial A}{\partial x} + \frac{\partial B}{\partial y} + \frac{\partial C}{\partial z} = 0. \quad (2)$$

The three functions A, B, C of the coordinates x, y, z must therefore be connected by this partial differential equation, which is also the condition (not only necessary, but also sufficient) for the definition of a function of a line $F[L]$ as the flux of the vector (A, B, C) connected with the line itself.

62. We shall now study more particularly the properties of those functions of lines $F[L]$ for which the quantities A, B, C can be chosen independent of the curve L. Given two curves

---

([1]) Cf. VOLTERRA, (13).

L_1, L_2 with a common segment taken in opposite senses, we shall call the curve formed by combining L_1 and L_2, after suppressing the common segment, the sum-curve of the two ($L_1 + L_2$). If we denote by σ_1 and σ_2 two portions of surface bounded by L_1 and L_2 respectively, and by $\sigma_1 + \sigma_2$ the surface obtained by combining σ_1 and σ_2, and bounded by $L_1 + L_2$, we shall have, by formula (6) of No. 60,

$$F[L_1] = \int_{\sigma_1} (A \cos n\, x + B \cos n\, y + C \cos n\, z) \, d\sigma,$$

$$F[L_2] = \int_{\sigma_2} (A \cos n\, x + B \cos n\, y + C \cos n\, z) \, d\sigma,$$

$$F[L_1 + L_2] = \int_{\sigma_1 + \sigma_2} (A \cos n\, x + B \cos n\, y + C \cos n\, z) \, d\sigma,$$

and therefore

$$F[L_1 + L_2] = F[L_1] + F[L_2]. \tag{3}$$

This relation expresses the *additive property* of these functions of lines; we shall therefore say that they are *of the first degree* or *linear*, by analogy with ordinary linear functions, for which $f(x + y) = f(x) + f(y)$. It is to be noted that considered as *functionals* of the three functions $x(s)$, $y(s)$, $z(s)$ they would also be *linear functionals*, in the sense indicated above in No. 19, with respect to $x(s)$, $y(s)$, $z(s)$ separately, since, e.g., the functional derivative with respect to $x(s)$ at the point s is, by formula (4) of No. 60, $X = Bz'(s) - Cy'(s)$, and is no longer dependent on the function $x(s)$[1].

The theory of functions of lines of a degree above the first and the expansions in series of functions of lines of degrees indefinitely increasing have been treated by C. FABRI[2], with an extension to hyperspaces.

63. Next, in order to see the significance of the three functions A, B, C, we take any curve whatever L and make it

---

[1] Given the definition of Chapter I of a "functional of the first degree", we have in fact to show that the previous definition of a "function of a line of the first degree" is not in contradiction with it, but rather, in conformity with footnote [2] to No. 58, is a particular case of it. This is in fact so, for if the first derivative of a functional does not depend on the function that is considered the independent variable (i.e. is a functional of degree zero), the functional is certainly linear, and also regular (No. 34).

[2] Cf. FABRI, (19) and (20) of Bibliography I.

vary in the neighbourhood of one of its points, and we shall denote by L_1 the curve obtained by this variation, and by Σ the surface generated by L during the variation from L to L_1. The corresponding increment $\Delta\mathrm{F} = \mathrm{F}[L_1] - \mathrm{F}[L]$ of the function of the line will be, by (6) of No. 60,

$$\Delta F = \int_{\Sigma} (A \cos n\, x + B \cos n\, y + C \cos n\, z)\, d\sigma;$$

and if we make the surface Σ decrease indefinitely and tend to 0 around the point considered in such a way that it comes to coincide with an element of a plane normal to the direction n, we shall have

$$\Delta F = (A \cos n\, x + B \cos n\, y + C \cos n\, z)\, \Sigma + \varepsilon,$$

where ε is an infinitesimal of higher order than Σ, and, in the limit,

$$\lim_{\Sigma = 0} \frac{\Delta F}{\Sigma} = A \cos n\, x + B \cos n\, y + C \cos n\, z = \frac{dF}{d\sigma}. \quad (4)$$

The limit $\dfrac{dF}{d\sigma}$ of the incremental ratio $\dfrac{\Delta F}{\Sigma}$, when the small area Σ tends to 0 in such a way that the directions of the normals at all the points of Σ differ by infinitely little from the direction n of the normal to a certain plane, will therefore be called the *derivative of the function of a line* $\mathrm{F}[L]$ *with respect to this plane at the point in question*. If, in particular, the plane is one of the coordinate planes yz, zx, xy, we shall have, for the three derivatives $\dfrac{dF}{d(yz)}$, $\dfrac{dF}{d(zx)}$, $\dfrac{dF}{d(xy)}$ with respect to these planes at a certain point,

$$\frac{dF}{d(yz)} = A, \quad \frac{dF}{d(zx)} = B, \quad \frac{dF}{d(xy)} = C. \quad (5)$$

We see therefore that *the three functions A, B, C of the point in space are precisely the three derivatives of the function $\mathrm{F}[L]$ with respect to the three coordinate planes*. We have also the further important property for these functions of lines of the first degree (for which A, B, C are independent of the curve considered) that *the derivative with respect to any plane whatever* at a certain point depends, by (4), *only on this point and on the orientation of the plane, and not on the curve.*

A third property of these functions of lines, lastly, can be deduced from (6) of No. 60, by reversing the sense of the curve. If in fact we denote by $-L$ the curve L described in the opposite sense, we shall have $F[-L] = -F[L]$, since this is equivalent to reversing the sense of the normal n to the surface σ bounded by L.

64. Comparing the results now obtained for functions of lines with the well-known results for ordinary functions of a point $f(x, y, z)$, we see therefore that while for the latter we have several derivatives depending on the various *directions*, and in particular the three fundamental ones

$$\frac{\partial f}{\partial x} = X_1, \; \frac{\partial f}{\partial y} = X_2, \; \frac{\partial f}{\partial z} = X_3, \qquad (6)$$

taken in the directions of the three coordinate axes, for the functions of lines we have also several derivatives, but depending on the *positions* of planes, and in particular the three fundamental ones

$$\frac{dF}{d(yz)} = X_{23}, \; \frac{dF}{d(zx)} = X_{31}, \; \frac{dF}{d(xy)} = X_{12}, \qquad (7)$$

taken with respect to the three coordinate planes. When we know the three derivatives (6) of the function $f(x, y, z)$, the difference $f_1 - f_0$ of this function for two different points P_1 and P_0 is given by

$$f_1 - f_0 = \int_l \left(\frac{\partial f}{\partial x} dx + \frac{\partial f}{\partial y} dy + \frac{\partial f}{\partial z} dz \right), \qquad (8)$$

where l is any line whatever joining P_0 and P_1. Similarly, if we know the three derivatives (7) of a function of a line $F[L]$, then, noting that $\cos nx \, d\sigma = dy \, dz$, $\cos ny \, d\sigma = dz \, dx$, $\cos nz \, d\sigma = dx \, dy$, the difference $F[L_1] - F[L_0]$ of the function of the line will, by (3), be given by

$$F[L_1] - F[L_0] = \int_\sigma \left(\frac{dF}{d(yz)} dy \, dz + \frac{dF}{d(zx)} dz \, dx + \frac{dF}{d(xy)} dx \, dy \right), \qquad (9)$$

where σ is any surface whatever having L_1 and L_0 for its complete boundary (generated by a closed curve that moves continuously from the initial position L_0 to the final position L_1). So also, while for (6) the conditions of integrability are given by

$$\frac{\partial X_3}{\partial y} = \frac{\partial X_2}{\partial z}, \ \frac{\partial X_1}{\partial z} = \frac{\partial X_3}{\partial x}, \ \frac{\partial X_2}{\partial x} = \frac{\partial X_1}{\partial y}, \tag{10}$$

for (7), on the contrary, the only condition, which we can also call the *condition of integrability*, is given by (2) of No. 61, and is therefore

$$\frac{\partial X_{23}}{\partial x} + \frac{\partial X_{31}}{\partial y} + \frac{\partial X_{12}}{\partial z} = 0. \tag{11}$$

§ 3. *Connectivity of Spaces in relation to the Polydromy of Functions of Lines of the First Degree.*

65. In trying to determine the exact significance of the formulae (8) and (9) of the preceding section, we have next to consider the various possible types of connectivity of the region R of space in which our functions are defined and regular. Thus, for the case of a perfect differential $\frac{\partial f}{\partial x}dx + \frac{\partial f}{\partial y}dy + \frac{\partial f}{\partial z}dz$, we can say that its integral from a point P_0 to a point P_1 has in general a value independent of the path l taken to go from P_0 to P_1 only if l is replaced by another line l', also joining P_0 and P_1, which can be obtained from l by making l vary continuously and without ever going outside the region considered. In other terms, the integral $\int_l df$ will be equal to $\int_{l'} df$ only if the *closed curve* consisting of l and l' together is reducible to a point by continuous deformation and without ever going outside the region considered; or, as it can also be expressed, if the region has *simple linear connectivity*. Examples of regions with simple linear connectivity are the space inside a sphere or a parallelopiped, or between two concentric spheres. For all these regions, the integral of a perfect differential is completely defined at every point when its value is given at any one point; i.e. it is a *monodromic* or *single-valued* func-

III ANALYTIC FUNCTIONS

tion of the point. This is no longer the case, in general, if we take, e.g., the region of space *inside a tore*, since there evidently exist within it closed curves C that are not reducible to a point without going outside the tore itself (a region with *multiple linear connectivity*); and if we take two points A and B on one of these curves C, the integral $\int_A^B df$ will in general have different values according as the path chosen to go from A to B follows one or the other of the two arcs into which the two points divide the curve. More precisely, the difference between the two values of the integral will be exactly $\int_C df$, which is called the *modulus of periodicity*; and the value of the integral itself at any point whatever, when its value at a particular point is given, will not in general be completely determined, but can vary by multiples of the modulus of periodicity (the multipliers being arbitrary integers), unless the modulus is 0. Hence if in a region with multiple linear connectivity the various moduli of periodicity are not all zero, the integral of a perfect differential defined within the region will have at every point a series of different values according to the path followed to reach that point; i.e. it will be a *polydromic* or *many-valued* function of the point.

66. We find precisely similar results for our functions of lines, the only change being the substitution *of superficial connectivity* of the region for *linear connectivity*. Formula (9) of No. 64, in the case when the initial curve L_0 reduces to a point, gives the value $F[L]$ of the function of the line in the form

$$F[L] = \int_\sigma \left(\frac{dF}{d(yz)} dy\, dz + \frac{dF}{d(zx)} dz\, dx + \frac{dF}{d(xy)} dx\, dy \right), \quad (1)$$

where σ is a surface having L for its complete boundary; and if σ' is another surface also having L for its boundary, and obtainable from σ by continuous deformation without ever going outside the region considered, then $\int_\sigma$ will be equal to $\int_{\sigma'}$, since, by the condition of integrability (11) of No. 64, the integral taken over the closed surface consisting of σ and σ' together (changing on one of them the sense of the normal) is zero. We can also say that in (1) σ can be replaced by another surface σ' which also has the curve L for its complete boundary, provided that the closed surface consisting of σ and σ' is reducible to a point by continuous deformation, without ever going outside the region considered. In particular, this will always be possible when this region has *simple superficial connectivity*, i.e. when *any* closed surface *whatever* is reducible to a point. Examples of regions with this kind of connectivity are the region inside a sphere or a parallelopiped, etc.

If now we take a region of space with multiple superficial connectivity, matters are quite otherwise. Let us take, for example, the region between two concentric spheres, which, as we have seen above, has *simple linear connectivity*; it has, on the contrary, *multiple superficial connectivity*, since there exist inside it closed surfaces Σ (e.g. the concentric spherical surfaces lying between the two that form the boundary of the region) which are not reducible to points by continuous deformation without going outside the region in question. If we take the simple curve L on one of these closed surfaces Σ, the surface will be divided into two portions or caps, which we shall denote by σ and σ' respectively, each bounded by the same curve L. We shall therefore have two different expressions for our function of the line $F[L]$: one as an integral taken over σ (cf. formula (1)), and the other as an integral of the same expression taken over σ'; and *if the integral taken over the closed surface Σ composed of σ and σ' together is not zero* (as will in general be the case, since Σ cannot be reduced to a point), *these two expressions will give distinct values* for $F[L]$. The difference between the two values will be precisely $\int_\Sigma$, which will also be called the *modulus of periodicity* of our function of the line; this function, being many-valued, will be called *polydromic*. In general therefore, the different values of a polydromic function of a line taken for one particular curve L will depend on the way in which, starting from a point and expanding and contracting a closed curve, the surface σ has been described which has for its boundary L (the final position of the moving curve); and these values will all differ by integral multiples of the moduli of periodicity, given by the values of the integrals $\int_\Sigma$ taken over those closed surfaces Σ which, for the particular type of superficial connectivity of the region, are not reducible to a point by continuous deformation.

We thus see that just as for integrals of perfect differentials (integrals taken along *lines*) the polydromy (or many-valuedness) is closely connected with the *linear connectivity* of the region, and the moduli of periodicity are given by the integrals taken along those closed curves which are not reducible to points (one-dimensional cycles), so for functions of lines of the first degree (integrals taken over *surfaces*) the polydromy is closely connected with the *superficial connectivity* of the region itself, and the moduli of periodicity are given by the integrals taken over those closed surfaces which are not reducible to points by continuous deformation (two-dimensional cycles).

67. The consideration of three-dimensional spaces, either simply or multiply connected, which are not limited by any bounding surfaces would seem to follow on naturally here. For an intuitive representation of these spaces VOLTERRA has devised a series of plastic models, photographs of which are reproduced at the end of LEFSCHETZ's book on *Analysis Situs*([1]), with the necessary explanations.

§ 4.

68. The foregoing considerations on functions of lines of the first degree, relating to three-dimensional regions, can be extended to n-dimensional regions; and if $n > 3$ we could also, in addition to the functions of points and lines, consider functions of (closed) surfaces, etc., or in general functions of (closed) hyperspaces of $1, 2, \ldots n - 2$ dimensions. The functions $F[S_i]$ of a hyperspace S_i of i dimensions ($i = 1, 2, \ldots n - 2$) of the first degree will be expressed by integrals taken through regions S_{i+1} of $i + 1$ dimensions which have the closed hyperspace S_i for their complete boundary ([2]). For these functions of hyperspaces, too, the hyperspace S_{i+1} of integration can be replaced by any other S'_{i+1} whatever which has the same boundary and can be obtained from the first by continuous deformation; but as the study of the polydromy of these functions $F[S_i]$ of hyperspaces is intimately connected with the $(i + 1)$-dimensional connectivity of the region considered, it becomes steadily more complicated as the number of dimensions increases.

Another possible extension would be, while still considering, e.g., functions of lines, to make the number n of the dimensions of the space in which the lines are immersed tend to infinity,

---

([1]) Cf. LEFSCHETZ, (5).
([2]) Cf. VOLTERRA, (14), (19), (21).

and so replace the discontinuous index i of the n coordinates $x_i(s)$ $(i = 1, 2, \ldots n)$ of the points of the line by a continuous parameter λ, according to the usual method. The functions of lines $F[L]$ in these spaces with an infinite number of dimensions would thus become functionals, not of the n functions $x_i(s)$ of one variable, but of a single function $x(\lambda, s)$ of two variables. These functions of lines have not yet been studied, and the attention of investigators may therefore be called to them.

§ 5.

69. Returning to three-dimensional space we shall next examine the two ways of extending to this space the ordinary theory of functions of a complex variable $z = x + \iota y$ (represented by the points x, y of a plane). A first method capable of extension is that given by RIEMANN's theory of *conjugate functions*; we can say, in fact, that the quantity $u + \iota v = f(x + \iota y)$ is a function of the complex variable $x + \iota y$ if the two real functions $u = u(x, y)$, $v = v(x, y)$ are conjugate, i.e. if they satisfy the conditions

$$\frac{\partial u}{\partial x} = \frac{\partial v}{\partial y}, \qquad \frac{\partial u}{\partial y} = -\frac{\partial v}{\partial x},$$

or, which is the same thing,

$$\frac{\partial u}{\partial x} = \frac{\partial v}{\partial y}, \qquad \frac{\partial u}{\partial y} = \frac{\partial v}{\partial (-x)}. \tag{1}$$

These conditions are contained in another more general one, itself a consequence of these two, which can be expressed by saying that if s and n are two directions normal to each other and forming, in this order, a pair congruent to the pair x, y, of the two co-ordinate axes, then for the two functions $u(x, y)$, $v(x, y)$ of a point defined in the plane x, y we must have

$$\frac{du}{ds} = \frac{dv}{dn}. \tag{2}$$

As a consequence we have further that two conjugate functions u and v (i.e. which satisfy (1) and (2)) also satisfy LAPLACE's equation

$$\frac{\partial^2 u}{\partial x^2} + \frac{\partial^2 u}{\partial y^2} = 0, \qquad \frac{\partial^2 v}{\partial x^2} + \frac{\partial^2 v}{\partial y^2} = 0, \qquad (3)$$

or, more shortly, $\Delta^2 u = 0$, $\Delta^2 v = 0$. Two conjugate functions are in consequence always *harmonic functions*. Vice versa, given a harmonic function, i.e. a solution of (3), there always exists another harmonic function *conjugate* to it.

A second method is that of CAUCHY, based on *monogeneity*. Given two complex quantities $w = u + \iota v$, $z = x + \iota y$, with $u = u(x, y)$, $v = v(x, y)$, which can be made to correspond to a *point* x, y, of a *plane*, we say that w is a *monogenic* function of z if the limit $\dfrac{dw}{dz}$ of the ratio $\dfrac{\Delta w}{\Delta z}$ of the increments is *independent of the direction* in which the point $x + \Delta x$, $y + \Delta y$ is made to tend towards the initial point x, y.

For ordinary functions of two variables, i.e. for *functions of a point* in the *plane*, the two methods are equivalent, for if u and v are conjugate, $w = u + \iota v$ is a monogenic function of $z = x + \iota y$, and *vice versa*. But when we pass from the plane to three-dimensional space the two methods diverge and we get two different generalisations of the primitive concept of an analytic function.

70. *Conjugate functions in space.* We shall begin with the theory which arises out of the first of these generalisations, that of *conjugate functions* in space. Starting from condition (2) of No. 69, which holds for the plane, let us take, instead of a pair, a set of three orthogonal directions n_1, n_2, n_3, congruent to the set of three directions x, y, z of the coordinate axes. To define a pair of conjugate functions in space, we shall now have to consider not, as in (2) of No. 69, two derivatives in two perpendicular directions, but instead one derivative in *one direction*, e.g. n_1, and one with respect to a plane $n_2 n_3$ normal to n_1, and having the same relation to n_1 as the plane yz has to the positive direction of x; we have therefore no longer two functions of points, but one function of a point (which has derivatives in the various directions at the point) and one of a line (which has derivatives with respect to the various planes through a point on the line).

We shall therefore say that a function $f(x, y, z)$ of a point and a function $F[L]$ of a line of the first degree are mutually *conjugate* when the relation

$$\frac{df}{dn} = \frac{dF}{d\sigma} \tag{4}$$

between their derivatives holds for every point, where n is any direction whatever, and $d\sigma$ lies in a plane normal to n and having the same relation to n as the plane yz has to the x-axis. From (4) we get as particular cases

$$\left.\begin{aligned}\frac{\partial f}{\partial x} &= \frac{dF}{d(yz)} \\ \frac{\partial f}{\partial y} &= \frac{dF}{d(zx)} \\ \frac{\partial f}{\partial z} &= \frac{dF}{d(xy)}\end{aligned}\right\}, \tag{5}$$

conditions which collectively are equivalent to (4). From these, and using condition (11) of No. 64, which must be satisfied by the three derivatives with respect to the three coordinate planes of a function of a line (condition of integrability), we get the result that if $f(x, y, z)$ can have a conjugate function $F[L]$, it must satisfy LAPLACE's equation

$$\Delta^2 f = \frac{\partial^2 f}{\partial x^2} + \frac{\partial^2 f}{\partial y^2} + \frac{\partial^2 f}{\partial z^2} = 0, \tag{6}$$

i.e. it must be *harmonic*; and *vice versa*, for every harmonic function $f(x, y, z)$ it will always be possible to find a function $F[L]$ of a line conjugate to it, defined by the equations (5).

So also, if, given a function $F[L]$ of a line there exists a function $f(x, y, z)$ of a point conjugate to it, we shall have by (5) and by the conditions of integrability (10) of No. 64 that the relation

$$\frac{\partial}{\partial y}\frac{dF}{d(yz)} = \frac{\partial}{\partial x}\frac{dF}{d(zx)}$$

and others similar to it must hold between the derivatives of $F[L]$ with respect to the three coordinate planes; or changing the sense of the plane zx to xz, so that the corresponding derivative changes sign,

$$\left.\begin{array}{l}\dfrac{\partial}{\partial x}\dfrac{d\mathrm{F}}{d(xz)}+\dfrac{\partial}{\partial y}\dfrac{d\mathrm{F}}{d(yz)}=0\\[2pt]\dfrac{\partial}{\partial y}\dfrac{d\mathrm{F}}{d(yx)}+\dfrac{\partial}{\partial z}\dfrac{d\mathrm{F}}{d(zx)}=0\\[2pt]\dfrac{\partial}{\partial z}\dfrac{d\mathrm{F}}{d(zy)}+\dfrac{\partial}{\partial x}\dfrac{d\mathrm{F}}{d(xy)}=0\end{array}\right\}. \qquad (7)$$

If we apply the term *harmonic* also to those functions of lines of the first degree which with their derivatives satisfy the differential system (7), we can say that *the necessary and sufficient condition for the existence of a function of a point conjugate to a given function of a line is that this function of a line shall be harmonic.*

From the above considerations, the theory of conjugate functions in three-dimensional space thus comes to be intimately connected with the theory of harmonic functions, which is a highly developed branch of mathematical science. Here all that can be done is to refer the reader to GAUSS's theorem and GREEN's theorem for the solution of LAPLACE's equation ([1]).

71. Examples of conjugate functions in space occur in physics when we are considering the potential $f(x, y, z)$ of some magnetic masses with respect to unit magnetic pole situated at the point whose coordinates are x, y, z (a function of a point), or the potential of the same masses with respect to a current of unit intensity flowing in a circuit in the form of a closed curve L (a function of the line L). It is easy to see that the function of a point given by the potential with respect to a pole is *conjugate*, in the sense described above, to the function of a line given by the potential with respect to the current.

72. Having thus established the theory of conjugate functions for space of three dimensions, we can pass to the consideration of conjugate functions in spaces of a greater number of dimensions. If n denotes this number, then to every function $\mathrm{F}[S_i]$ of a hyperspace ($i = 1, 2, \ldots n - 2$) can be associated, under suitable hypotheses (analogous to the conditions (6) and (7) of No. 70) another function $\mathrm{F}[S_{n-i-2}]$ of a hyperspace conjugate to the first ([2]). In order that these two functions may be of the same kind, i.e. that the hyperspaces S_i and S_{n-i-2} may have the same dimensions, we must have

---

([1]) Cf. VOLTERRA, (15), (16), (17).
([2]) Ibid.

$i = n - i - 2$, or $n = 2(i + 1)$. If the number n of dimensions is even, it follows that there exist pairs of mutually conjugate functions of hyperspaces $F\left[S_{\frac{n}{2}-1}\right]$ of the same kind. Thus for $n = 2$ (in the plane), we have pairs of conjugate functions of points $F[S_0] = f(x, y)$; for $n = 4$, we have pairs of mutually conjugate functions of lines $F[S_1] = F[L]$; etc.

§ 6. *Monogeneity and Isogeneity.*

73. Passing now to the extension of the concept of monogeneity, we may remind the reader that two complex functions $\varphi = \varphi_1 + \iota \varphi_2$, $f = f_1 + \iota f_2$ of the *points* of a *surface* are said to be *monogenic* functions of each other when the differential quotient $\dfrac{d\varphi}{df} = \lim \dfrac{\Delta \varphi}{\Delta f}$ is independent of the direction in which the variable point (to which $\varphi + \Delta \varphi$, $f + \Delta f$ correspond) tends to the initial point (to which φ and f correspond). If u and v denote the curvilinear co-ordinates of the surface, this condition is equivalent to

$$\frac{\dfrac{\partial (\varphi_1 + \iota \varphi_2)}{\partial u}}{\dfrac{\partial (f_1 + \iota f_2)}{\partial u}} = \frac{\dfrac{\partial (\varphi_1 + \iota \varphi_2)}{\partial v}}{\dfrac{\partial (f_1 + \iota f_2)}{\partial v}}. \tag{1}$$

Putting

$$\left. \begin{array}{c} \dfrac{\partial f_1}{\partial u} = p_1, \quad \dfrac{\partial f_1}{\partial v} = q_1, \quad \dfrac{\partial f_2}{\partial u} = p_2, \quad \dfrac{\partial f_2}{\partial v} = q_2, \\ p_1^2 + p_2^2 = \varepsilon_{11}, \quad q_1^2 + q_2^2 = \varepsilon_{22}, \quad p_1 q_1 + p_2 q_2 = \varepsilon_{12}, \\ p_2 q_1 - p_1 q_2 = D, \end{array} \right\} \tag{2}$$

the condition (1) is transformed into the following two:

$$\left. \begin{array}{c} \dfrac{\partial \varphi_2}{\partial u} = \dfrac{\varepsilon_{11} \dfrac{\partial \varphi_1}{\partial v} - \varepsilon_{12} \dfrac{\partial \varphi_1}{\partial u}}{D} \\ \dfrac{\partial \varphi_2}{\partial v} = \dfrac{-\varepsilon_{22} \dfrac{\partial \varphi_1}{\partial u} + \varepsilon_{12} \dfrac{\partial \varphi_1}{\partial v}}{D} \end{array} \right\}, \tag{3}$$

connecting the derivatives of φ_1 and φ_2; further, both φ_1 and φ_2 satisfy LAPLACE's equation.

74. Now let us consider two complex functions of lines

$\Phi[L] = \Phi_1 + \iota\Phi_2$ and $F[L] = F_1 + \iota F_2$, and let us calculate the derivatives of these two functions of the same line L at the same point and with respect to the same plane. If their ratio

$$\frac{\dfrac{d\Phi}{d\sigma}}{\dfrac{dF}{d\sigma}} = \frac{d\Phi}{dF} \tag{4}$$

depends only on the point and *does not depend on the plane* in which σ lies, then the two functions $\Phi[L]$ and $F[L]$ are said to be *isogenic* functions of each other; and the differential ratio $\dfrac{d\Phi}{dF} = f(x, y, z)$ (a function of a point) is called the *derivative* of Φ with respect to F at the point considered. As for monogenic functions of points, so for isogenic functions of lines we get relations, analogous to (3), between the derivatives with respect to the three co-ordinate planes. More precisely, putting

$$\left. \begin{array}{l} \dfrac{dF_1}{d(yz)} = p_1, \quad \dfrac{dF_1}{d(zx)} = q_1, \quad \dfrac{dF_1}{d(xy)} = r_1 \\[1ex] \dfrac{dF_2}{d(yz)} = p_2, \quad \dfrac{dF_2}{d(zx)} = q_2, \quad \dfrac{dF_2}{d(xy)} = r_2 \\[1ex] \varepsilon_{23} = q_1 r_1 + q_2 r_2, \quad D_1 = q_2 r_1 - q_1 r_2 \\[1ex] \varepsilon_{31} = r_1 p_1 + r_2 p_2, \quad D_2 = r_2 p_1 - r_1 p_2 \\[1ex] \varepsilon_{12} = p_1 q_1 + p_2 q_2, \quad D_3 = p_2 q_1 - p_1 q_2 \end{array} \right\}, \tag{5}$$

we get from (4), by considering the derivatives with respect to the three co-ordinate planes, the three equivalent relations [1]

$$\left. \begin{array}{l} \dfrac{d\Phi_2}{d(yz)} = \dfrac{\varepsilon_{12}\dfrac{d\Phi_1}{d(xy)} - \varepsilon_{13}\dfrac{d\Phi_1}{d(xz)}}{D_1} \\[3ex] \dfrac{d\Phi_2}{d(zx)} = \dfrac{\varepsilon_{23}\dfrac{d\Phi_1}{d(yz)} - \varepsilon_{21}\dfrac{d\Phi_1}{d(yx)}}{D_2} \\[3ex] \dfrac{d\Phi_2}{d(xy)} = \dfrac{\varepsilon_{31}\dfrac{d\Phi_1}{d(zx)} - \varepsilon_{32}\dfrac{d\Phi_1}{d(zy)}}{D_3} \end{array} \right\}, \tag{6}$$

---

[1] Cf. VOLTERRA, (13), (18); EVANS, (17) of Bibliography I, Lecture II.

which, if the derivatives of F, and therefore the ε_{hk}'s ($h, k = 1, 2, 3$) and the D_h's ($h = 1, 2, 3$), are given, connect the derivatives of the two functions of lines $\Phi_1[L]$ and $\Phi_2[L]$, the real part and the coefficient of the imaginary part of the function Φ which is isogenic to F. From (6) we get further

$$D_1 \frac{d\Phi_2}{d(yz)} + D_2 \frac{d\Phi_2}{d(zx)} + D_3 \frac{d\Phi_2}{d(xy)} = 0$$

and similarly for Φ_1.

From the equations (6) and the condition of integrability (11) of No. 64, there follows another equation, analogous to LAPLACE's equation, *connecting the derivatives of* Φ_1, and similarly for Φ_2. For the functions of lines which satisfy this differential equation, various characteristic properties can be established analogous to those of harmonic functions.

75. We shall now show that two functions which are isogenic to a third are isogenic to each other.

Let Φ be isogenic to F; this means that the ratio of the two derivatives $\dfrac{d\Phi}{d\sigma} : \dfrac{dF}{d\sigma} = f$ is independent of the plane with respect to which they are calculated; the same holds for the ratio $\dfrac{d\Psi}{d\sigma} : \dfrac{dF}{d\sigma} = f_1$ if Ψ is also isogenic to F. But it follows that the ratio

$$\frac{d\Psi}{d\sigma} : \frac{d\Phi}{d\sigma} = \frac{f_1}{f}$$

will also be independent of the plane with respect to which the derivatives are taken, which means precisely that Ψ is isogenic to Φ. We can therefore say that all the functions of lines which are isogenic to a given one form a *group* of functions which are all mutually isogenic.

76. We have already seen in No. 74 that by the very definition of isogeneity the derivative $\dfrac{d\Phi}{dF} = f(x, y, z)$ of a function of a line Φ with respect to another F, isogenic to it, is a function only of the point x, y, z ([1]). Remembering that, also by

---

([1]) Cf. VOLTERRA, (13), (18).

ANALYTIC FUNCTIONS

the definition of isogeneity, we have

$$\frac{d\Phi}{d(yz)} = f\frac{dF}{d(yz)}, \quad \frac{d\Phi}{d(zx)} = f\frac{dF}{d(zx)}, \quad \frac{d\Phi}{d(xy)} = f\frac{dF}{d(xy)},$$

and remembering the condition of integrability (11) of No. 64, we reach the result that *the function of a point f must satisfy the linear homogeneous partial differential equation of the first order*

$$\frac{\partial f}{\partial x}\frac{dF}{d(yz)} + \frac{\partial f}{\partial y}\frac{dF}{d(zx)} + \frac{\partial f}{\partial z}\frac{dF}{d(xy)} = 0. \qquad (7)$$

A function of a point f which satisfies a differential equation of this kind is said to be *isogenic to the function of a line* F; it will be seen at once that it is also isogenic to all other functions of lines which are isogenic to F.

We have seen that the derivative f of a function of a line Φ with respect to another F, isogenic to it, satisfies equation (7) (is isogenic to the two functions F and Φ). *Vice versa*, if a function of a point f satisfies (7), we shall also have, by the condition of integrability that is satisfied by the three derivatives of F with respect to the three co-ordinate planes,

$$\frac{\partial}{\partial x}\left(f\frac{dF}{d(yz)}\right) + \frac{\partial}{\partial y}\left(f\frac{dF}{d(zx)}\right) + \frac{\partial}{\partial z}\left(f\frac{dF}{d(xy)}\right) = 0, \qquad (8)$$

which is therefore the necessary and sufficient condition (condition of integrability) for the existence of a function of a line Φ which shall have for its derivatives

$$\frac{d\Phi}{d(yz)} = f\frac{dF}{d(yz)}, \quad \frac{d\Phi}{d(zx)} = f\frac{dF}{d(zx)}, \quad \frac{d\Phi}{d(xy)} = f\frac{dF}{d(xy)}. \qquad (9)$$

Hence, *if a function of a point f satisfies* (7), *i.e. is isogenic to the function of a line* F, *there certainly exists another function of a line* Φ, *defined by the equations* (9), *isogenic to* F, *and such that its derivative* $\dfrac{d\Phi}{dF}$ *with respect to* F *is precisely the function of a point* $f(x, y, z)$; Φ will then be called the integral of f, and will be written

$$\Phi[L] = \int f\,dF = \int_\sigma \left[f\frac{dF}{d(yz)}\,dy\,dz + f\frac{dF}{d(zx)}\,dz\,dx \right.$$
$$\left. + f\frac{dF}{d(xy)}\,dx\,dy \right],$$

where σ is a surface bounded by the closed curve L. To obtain Φ it is therefore necessary to calculate a surface integral.

77. Next let us consider several functions of points $f, f_1, f_2, \ldots$, isogenic to a single function of a line $F[L]$; we shall say that they are mutually *isogenic*. Since they must all satisfy equation (7) of No. 76, if we take any three of them, f, f_1, f_2, and if the three derivatives of F do not vanish identically, then the functional determinant (Jacobian) $\dfrac{d(f, f_1, f_2)}{d(x, y, z)}$ must be zero; or, in other words, one of the f's must be a function of the other two, say, $f = \varphi(f_1, f_2)$. Vice versa, if f_1 and f_2 are isogenic to a function of a line F (i.e. satisfy (7) of No. 76), then any other function $f = \varphi(f_1, f_2)$ of f_1 and f_2 will also be isogenic to F. We can therefore say also that the aggregate of all the functions of points f isogenic to a given function of a line is obtained from any two of them f_1 and f_2 (independent of each other) by taking all the functions $f = \varphi(f_1, f_2)$ of these two.

Given the function of a line $F[L]$, we have thus set out the properties of the functions of points isogenic to it and a method of constructing all these functions. If instead we start from any two functions of points whatever, f_1 and f_2, we have identically

$$\frac{\partial}{\partial x}\frac{d(f_1, f_2)}{d(y, z)} + \frac{\partial}{\partial y}\frac{d(f_1, f_2)}{d(z, x)} + \frac{\partial}{\partial z}\frac{d(f_1, f_2)}{d(x, y)} = 0 \qquad (10)$$

$\left(\text{where } \dfrac{d(f_1, f_2)}{d(y, z)}\right.$ denotes the Jacobian determinant of f_1 and f_2 with respect to y and z, etc.); hence, by No. 64, there exists a function of a line F such that

$$\frac{dF}{d(yz)} = \frac{d(f_1, f_2)}{d(y, z)}, \quad \frac{dF}{d(zx)} = \frac{d(f_1, f_2)}{d(z, x)}, \quad \frac{dF}{d(xy)} = \frac{d(f_1, f_2)}{d(x, y)}. \quad (11)$$

This function of a line $F[L]$ will evidently be isogenic to both f_1 and f_2, and to all the functions of points $f = \varphi(f_1, f_2)$ isogenic to them; it will also be denoted by $(f_1, f_2) = F[L]$, while its differential dF will be denoted by $d(f_1, f_2)$ [1].

78. We have already seen in No. 76 that if Φ is any function of a line whatever isogenic to F, and f its derivative $\dfrac{d\Phi}{dF}$,

---

[1] Cf. VOLTERRA, (13), (18).

then the function Φ is obtained by means of a surface integral

$$\Phi[L] = \int f d\, \mathrm{F} = \int_\sigma f \left[\frac{d\,\mathrm{F}}{d(yz)} dy\, dz + \frac{d\,\mathrm{F}}{d(zx)} dz\, dx + \frac{d\,\mathrm{F}}{d(xy)} dx\, dy \right];$$

but since $f = \varphi(f_1, f_2)$, this integral reduces by (11) to

$$\Phi[L] = \int f d\, \mathrm{F} = \int \varphi(f_1, f_2)\, df_1\, df_2. \qquad (12)$$

We thus see also that *the double integration of a function of two complex variables is a problem belonging to this theory of isogenic functions of lines.*

From (12) we get in particular, if the double integration is taken over a *closed* surface σ which does not contain any singularities,

$$\int_\sigma f\, d\,\mathrm{F} = \int_\sigma \varphi(f_1, f_2)\, df_1\, df_2 = 0, \qquad (13)$$

a property which extends to functions of two complex variables CAUCHY's well-known theorem

$$\int_s f(z)\, dz = 0 \qquad (14)$$

for the integral of a function of a single complex variable taken round a closed curve s in the plane which does not contain any singularities. *Vice versa*, a theorem analogous to MORERA's [1] theorem also holds, namely, that *if* (13) *holds for any closed surface σ whatever, then f is a function of a point isogenic to the function of a line* F, and also $\varphi(f_1, f_2)$ is a function in the ordinary sense of the two complex variables f_1 and f_2.

The theory of *isogenic* functions can be extended to the case of functions of hyperspaces. We thus obtain a theory which includes the theories of repeated complex integrations[2].

§ 7.

79. We shall conclude this chapter by an important proposition due to Dr. MANDELBROJT [3]. We know that given a function $f(x)$ of a real variable defined for values of the variable

---

[1] Cf. MORERA, (214) of Bibliography VI.
[2] Cf. VOLTERRA, (14), (19), (21).
[3] Cf. MANDELBROJT, (6), (7).

represented by points on a *straight line*, then if $f(x)$ can be expanded in a TAYLOR series, the definition of the function can be extended also to complex values $x + \iota y$ represented by points in a region of the *plane x, y*; we say in this case that the function $f(x + \iota y)$ of the complex variable is the *analytical continuation* of the original function $f(x)$ of the real variable.

We shall now show analogously that isogenic functions of lines can be considered as the analytical continuation in space of monogenic functions of points on a surface. If in fact $\Phi[L]$ and $F[L]$ are two mutually isogenic functions of lines, let us take a surface σ, and to any point Q of σ let one and only one closed curve L through Q be made to correspond, but in such a way that all these ∞^2 closed curves have two by two a common segment AMB, so that the sense is determined on all these curves when it is given on one of them; let us suppose further that if Q_1 is another point on σ to which the curve L_1 corresponds, the segment AMB common to L and L_1 increases when Q_1 tends to Q, so that the two complementary segments AQB and AQ_1B of L and L_1 respectively both tend to zero. Now let us define on σ two functions of points φ and f, making the values $\varphi = \Phi[L]$ and $f = F[L]$ correspond at every point Q, where L is the closed curve corresponding to Q; since Φ and F are mutually isogenic, it follows that $\lim\limits_{L_1 = L} \dfrac{\Delta \Phi}{\Delta F}$ (where $\Delta \Phi = \Phi[L_1] - \Phi[L]$, $\Delta F = F[L_1] - F[L]$) exists and is independent of the way in which the curve L_1, which is variable in the neighbourhood of a point, tends to the original curve L. If in particular Q_1 tends to Q, the curve L_1 corresponding to Q_1, which differs from L (corresponding to Q) only in the neighbourhood of Q, will tend towards L, and

$$\lim_{Q_1 = Q} \frac{\Delta \varphi}{\Delta f} = \lim_{L_1 = L} \frac{\Delta \Phi}{\Delta F}$$

will be independent of the direction along which Q_1 approaches Q on the surface; which proves precisely that the two functions φ and f previously defined for the points of σ are mutually *monogenic*. The two isogenic functions of lines Φ and F, which take the same values as φ and f for the particular ∞^2 curves L corresponding to the points of σ, can therefore be considered as the analytical continuation to curves in space of the two monogenic functions of points φ and f.

BIBLIOGRAPHY III.

(1) BOHANNAN (R. D.).—Pascal line equation and some consequences (*Math. Monthly*, Vol. XXIII, 1916.)
(2) DE DONDER.—Sur les fonctions de Volterra et les invariants intégraux. (*Publ. Acad. Royale de Belgique*, No. 6, June 1906.)
(3) FRÉCHET (M.).—Sur les fonctions de lignes fermées (*Ann. de l'École Norm.*, Vol. XXI, 3ième sér., 1904).
(4) GRAUSTEIN (W. C.).—Note on isogenous complex functions of curves. (*Bull. of Am. Math. Soc.*, Vol. 24, 1918.)
(5) LEFSCHETZ (S.).—L'analysis situs et la géométrie algébrique. (Coll. Borel.) Paris, Gauthier-Villars, 1924.
(6) MANDELBROJT (S.).—Sur le prolongement analytique des fonctions monogènes au sens de Cauchy en fonctions isogènes au sens de M. VOLTERRA. (*C. R. de l'Ac. des Sciences de Paris*, Vol. 180, 1er sem., 1925.)
(7) MANDELBROJT (S.).—Remarques sur la manière dont peuvent être engendrées les fonctionelles isogènes. (*C. R. de l'Ac. des Sciences de Paris*, Vol. 181, 2ème sem., 1925.)
(8) NICOLESCO (MIRON).—Fonctions complexes dans le plan et dans l'espace. (*Thèse de Doctorat présentée à la Faculté des Sciences de Paris*. Paris, Gauthier-Villars, 1928.)
(9) PASCAL (ERNESTO).—L'integrazione doppia nel campo complesso. (*Rend. Acc. Scienze di Napoli*, (3), Vol. XXIII, 1917.)
(10) PASCAL (ERNESTO).—Il teorema di Cauchy-Morera esteso agli integrali doppi delle funzioni di variabili complesse. (*Rend. Acc. Scienze di Napoli*, (3), Vol. XXIII, 1917.)
(11) PASCAL (MARIO).—Le funzioni monogenee di linee complesse. (*Rend. Acc. Scienze di Napoli*, (3), Vol. XXV, 1919).
(12) PASCAL (MARIO).—Il teorema e la formula di Cauchy per le funzioni monogenee di linee complesse. (*Rend. Acc. Scienze di Napoli*, (3), Vol. XXV, 1919.)
(13) VOLTERRA (V.).—Sopra un' estensione della teoria di Riemann sulle funzioni di variabili complesse. (*R. Acc. dei Lincei. Rend.*, ser. 4a, Vol. III, 1887; Vol. IV, 1888 (3 notes).)
(14) VOLTERRA (V.).—Delle variabili complesse negli iperspazi. (*R. Acc. dei Lincei. Rend.*, Vol. V, 1889 (2 notes).)
(15) VOLTERRA (V).—Sulle funzioni coniugate. (*R. Acc. dei Lincei. Rend.*, Vol. V, 1889.)
(16) VOLTERRA (V.).—Sulle funzioni di iperspazii e sui loro parametri differenziali. (*R. Acc. dei Lincei. Rend.*, Vol. V, 1889.)

(17) VOLTERRA (V.).—Sulla integrazione di un sistema di equazioni differenziali a derivate parziali che si presenta nella teoria delle funzioni coniugate. (*Rend. Circ. Mat. Palermo*, Vol. III, 1889.)

(18) VOLTERRA (V.)—Sur une généralisation de la théorie des fonctions d'une variable imaginaire. (*Act. Math. Stockholm*, 1889.)

(19) VOLTERRA (V.).—Sulle variabili complesse negli iperspazi. (*R. Acc. dei Lincei. Rend.*, Vol. VI, 1890.)

(20) VOLTERRA (V.).—Un teorema sugli integrali multipli. (*R. Acc. Sc. di Torino. Atti*, Vol. XXXII, 1897.)

(21) VOLTERRA (V.).—The generalization of analytic functions. (*Rice Inst. Pamph.*, Vol. III, No. 1, 1917.)

CHAPTER IV

THEORY OF COMPOSITION AND OF PERMUTABLE FUNCTIONS

§ 1. *Composition and Permutability of the First and Second Kinds.*

80. In Chapter II (No. 44) we have already seen examples of particular functionals $K^{(i)}(x,t)$ of functions of two variables $K^{(1)}(x,t)$, depending also on two parameters x, t (the *iterative kernels* of the kernel $K^{(1)}$, of the first and second kinds). In this chapter we propose to study in more detail an important class of functionals, including the foregoing as a particular case, which are of great utility for the solution of extensive classes of integral and integro-differential equations (cf. Nos 112,, 113, 114).

Given two functions $f(x,y)$, $g(x,y)$ of two variables, the function $h(x,y)$ defined by

$$h(x,y) = \int_x^y f(x,\xi) g(\xi,y) d\xi, \qquad (1)$$

which is a *functional of the two functions* f and g, will be called the *product by composition of the first kind* of the two functions f and g, and will be denoted by

$$h = \overset{**}{f g};$$

while the operation expressed by (1), by means of which we pass from the two functions f and g to the function h, will be called *composition of the first kind* of the two functions f and g.

Analogously, we shall say that $k(x,y)$ is the *product by composition of the second kind* of the two functions f and g when we have

$$k(x,y) = \int_a^b f(x,\xi) g(\xi,y) d\xi \qquad (2)$$

(where a, b are constant), and the operation expressed by (2) will be called *composition of the second kind*, and k will be denoted by the symbol

$$k = \overset{**}{f}\overset{**}{g}.$$

These operations of composition are an extension to the case of an infinite number of variables of the notion of the *product of two square matrices* $\|a_{ir}\|$, $\|b_{rs}\|$, or of the *composition of the corresponding linear substitutions* (i, r, $s = 1, 2, \ldots n$). Composition of the second kind corresponds to the case of general square matrices, while that of the first kind corresponds to the case of matrices $\|a_{ir}\|$ in which $a_{ir} = 0$ for $r \geq i$.

Corresponding to these two types of composition we have two different types of *permutability*. Two functions f and g are called *permutable of the first kind* if

$$\overset{**}{f g} = \overset{**}{g f},$$

i.e. if

$$\int_x^y f(x, \xi) g(\xi, y) \, d\xi = \int_x^y g(x, \xi) f(\xi, y) \, d\xi;$$

they are called *permutable of the second kind* if

$$\overset{**}{f}\overset{**}{g} = \overset{**}{g}\overset{**}{f},$$

i.e. if

$$\int_a^b f(x, \xi) g(\xi, y) \, d\xi = \int_a^b g(x, \xi) f(\xi, y) \, d\xi.$$

The products by composition of two, three or more permutable functions are permutable with one another and also with the given functions.

The operations of composition of the first and second kinds are evidently *associative*, i.e.

$$(\overset{**}{f g})\overset{*}{h} = \overset{*}{f}(\overset{**}{g h}),$$

$$(\overset{**}{f}\overset{**}{g})\overset{**}{h} = \overset{**}{f}(\overset{**}{g}\overset{**}{h}),$$

and *distributive*, i.e.

$$\overset{*}{f(g+h)} = \overset{**}{fg} + \overset{**}{fh},$$

$$\overset{**}{f(g+h)} = \overset{**}{f}\overset{**}{g} + \overset{**}{f}\overset{**}{h},$$

whatever the three functions f, g, h may be, while *they are not*, *in general, commutative.* If however we limit the investigation to aggregates of functions which are permutable two by two, the formal analogy between *composition* and *arithmetical product* will be perfect ([1]).

§ 2. Powers and Polynomials by Composition.

81. If a function $f(x, y)$ is compounded with itself, we shall have a function $\overset{**}{ff} = \overset{*}{f^2}$ which we shall call the *square by composition of the first kind* of f; and in general we shall call the function

$$\overset{*}{f^n} = \overset{*}{f^{n-1}} \overset{*}{f}$$

(n a positive integer) the nth *power by composition* of f.

Similarly for composition of the second kind.

The powers by composition of a function are permutable with one another and the rules of calculation for them are the same as those for ordinary powers.

If we consider an aggregate of permutable functions f_1, $f_2, \ldots f_i$, and form the sum of a finite number of terms of the type

$$a \overset{*}{f_1^{n_1}} \overset{*}{f_2^{n_2}} \ldots \overset{*}{f_i^{n_i}} \qquad (n_1, n_2, \ldots n_i \text{ positive integers})$$

obtained by the composition of several powers and multiplication by a constant, the expression so obtained will be called a *polynomial by composition of the first kind*; an analogous definition will hold for polynomials by composition of the *second kind*. For these polynomials by composition the following theorem holds:

A polynomial by composition constructed with permutable functions $f_1, f_2, \ldots$ *is a new function which is itself permutable with these functions; the composition of the polynomials will be carried out by the same rules as hold for the product of ordinary polynomials of the variables* $f_1, f_2, \ldots$ ([2]).

This algebra of permutable functions, which is perfectly analogous to ordinary algebra, was first studied by EVANS ([3]).

([1]) Cf. VOLTERRA and PÉRÈS, (44), p. 7. Evidently the product by composition can be equal to 0, while the factors are not equal to 0.
([2]) Cf. VOLTERRA and PÉRÈS, (44), p. 7.
([3]) Cf. G. C. EVANS, (7), (8), (9).

In addition to powers with a positive integral exponent, we also define, for convenience in calculating, a new and purely symbolic expression, which is called the power with exponent 0, by means of the formula

$$\overset{*}{f}{}^0 \overset{*}{g} = \overset{*}{g}\overset{*}{f}{}^0 = \overset{*}{g}, \qquad \overset{*}{g}{}^0 = \overset{*}{f}{}^0.$$

This symbol has the same function in the theory of composition as unity in the ordinary calculus. Fixing once for all the basic function, which has no effect, and taking it as 1, we shall often denote this symbol by $\overset{*}{1}{}^0 = \overset{*}{f}{}^0 = \overset{*}{g}{}^0$.

§ 3. *Series and Functions by Composition of the First Kind.*

82. Limiting our investigations for the present more especially to the theory of composition of the first kind, we can pass from polynomials by composition as defined in the preceding section to the consideration of more general functionals, obtained from ordinary power series by substituting *powers by composition* of a function f for *ordinary powers*.

Thus, for example, from the power series

$$z + z^2 + z^3 + \cdots + z^n + \cdots, \tag{1}$$

convergent within the circle $|z| < 1$, we can derive the other series

$$f + \overset{*}{f}{}^2 + \overset{*}{f}{}^3 + \cdots + \overset{*}{f}{}^n + \cdots, \tag{2}$$

which represents a functional that we have already considered (the solvent or reciprocal kernel of the kernel $-f(x,y)$; cf. No. 43). This series is *always convergent* whatever f may be, provided it is limited (cf. No. 43). This can also be expressed by saying that the series

$$zf + z^2 \overset{*}{f}{}^2 + z^3 \overset{*}{f}{}^3 + \cdots + z^n \overset{*}{f}{}^n + \cdots, \tag{3}$$

a functional of f, is, considered as a function of z, an *integral function*.

This property has the character of a general theorem which holds for any power series whatever (without a constant term, if we do not wish to adopt the symbolic expression $\overset{*}{f^0}$ defined above). In fact, if the series

$$a_1 z + a_2 z^2 + a_3 z^3 + \cdots + a_n z^n + \cdots \qquad (4)$$

is convergent for some value of z other than 0, this means that the sequence $\sqrt[n]{|a_n|}$ is limited, by the CAUCHY-HADAMARD theorem [1], and therefore $|a_n| < M^n$ (M a suitable constant). If then we construct the new series

$$a_1 z \overset{*}{f} + a_2 z^2 \overset{*}{f^2} + a_3 z^3 \overset{*}{f^3} + \cdots + a_n z^n \overset{*}{f^n} + \cdots, \qquad (5)$$

this will always be convergent, for any value of z, and for any f whatever provided it is limited, since if $|f(x, y)| < L$, we have also (No. 43)

$$|\overset{*}{f^n}| < \frac{L^n |y - x|^{n-1}}{(n-1)!},$$

and therefore

$$|a_n z^n \overset{*}{f^n}| < |MLz|^n \frac{|y - x|^{n-1}}{(n-1)!},$$

i.e. the terms of the series (5) are less in absolute value than the terms of an exponential series.

This theorem allows of yet a further extension. We can in fact show analogously that if

$$\varphi(z_1, z_2, \ldots z_n) = \sum_{1}^{\infty} {}_{i_1 i_2 \ldots i_n} a_{i_1 i_2 \ldots i_n} z_1^{i_1} z_2^{i_2} \ldots z_n^{i_n}$$

is a power series in the variables z_r which is convergent when the moduli $|z_r|$ ($r = 1, 2, \ldots n$) are sufficiently small, then the series

$$\sum_{i_1 i_2 \ldots i_n} a_{i_1 i_2 \ldots i_n} \overset{*}{f_1^{i_1}} \overset{*}{f_2^{i_2}} \ldots \overset{*}{f_n^{i_n}} \qquad (6)$$

(which we shall call a series by composition) is always convergent whatever the f_r's may be (provided they are limited), and represents

---

[1] Cf. A. CAUCHY: Cours d'Analyse de l'École Royale Polytechnique, première partie: Analyse Algébrique, p. 286; J. HADAMARD, (22) of Bibliography II, p. 17.
Cf. also L. BIANCHI: Funzioni di variabile complessa, p. 16. Pisa, Spoerri, 1916.

a *function of x and y which is permutable with all the f_r's if these are permutable with each other*.

The series (6) is evidently a *functional of the f_r's*, which we shall also call a *function by composition*, and shall denote by the symbol

$$\varphi(\overset{*}{f_1}, \overset{*}{f_2}, \ldots \overset{*}{f_n}).$$

We thus see how from any *analytic function* $\varphi(z_1, z_2, \ldots z_n)$, regular in the region round $z_1 = z_2 = \cdots = z_n = 0$, we can correlatively obtain a *functional* (function by composition). It is to be noted that for the product by composition of two functions by composition $\varphi_1(\overset{*}{f_1}, \overset{*}{f_2}, \ldots \overset{*}{f_n})$, $\varphi_2(\overset{*}{f_1}, \overset{*}{f_2}, \ldots \overset{*}{f_n})$, we have only to apply the rules that hold for the ordinary product of the two analytic functions $\varphi_1(z_1, z_2, \ldots z_n)$, $\varphi_2(z_1, z_2, \ldots z_n)$; in other words, if

$$\varphi_3(z_1, z_2, \ldots z_n) = \varphi_1(z_1, z_2, \ldots z_n)\varphi_2(z_1, z_2, \ldots z_n),$$

then the composition of $\varphi_1(\overset{*}{f_1}, \overset{*}{f_2}, \ldots \overset{*}{f_n})$ and $\varphi_2(\overset{*}{f_1}, \overset{*}{f_2}, \ldots \overset{*}{f_n})$ will give $\varphi_3(\overset{*}{f_1}, \overset{*}{f_2}, \overset{*}{f_3}, \ldots \overset{*}{f_n})$ as its result.

§ 4.

83. As an application, let us consider the series

$$e^{\lambda z} = 1 + \lambda z + \frac{\lambda^2}{2!}z^2 + \cdots + \frac{\lambda^n}{n!}z^n + \cdots$$

containing the parameter λ. For this function of z the formula

$$e^{\lambda z} e^{\mu z} = e^{(\lambda + \mu)z} \tag{7}$$

holds. Corresponding to this, if we construct the series by composition

$$u(\lambda; x, y) = \overset{*}{1^0} + \lambda \overset{*}{f} + \frac{\lambda^2}{2!}\overset{*}{f^2} + \cdots + \frac{\lambda^n}{n!}\overset{*}{f^n} + \cdots,$$

then it follows from the result stated above that the formula

$$\overset{*}{u}(\lambda; x, y)\overset{*}{u}(\mu; x, y) = \overset{*}{u}(\lambda + \mu; x, y) \tag{8}$$

will hold for it.

If we wish to avoid the introduction of the symbol $\overset{*}{1}{}^0$, we have only to consider the other series by composition (an integral function of λ and a functional of f)

$$v(\lambda;x,y) = \lambda f + \frac{\lambda^2}{2!}\overset{*}{f^2} + \cdots + \frac{\lambda^n}{n!}\overset{*}{f^n} + \cdots,$$

and substituting $\overset{*}{1}{}^0 + v$ for u in (8), we shall have

$$[\overset{*}{1}{}^0 + \overset{*}{v}(\lambda;x,y)][\overset{*}{1}{}^0 + \overset{*}{v}(\mu;x,y)] = \overset{*}{1}{}^0 + \overset{*}{v}(\lambda+\mu;x,y);$$

i.e.

$$v(\lambda+\mu;x,y) = v(\lambda;x,y) + v(\mu;x,y) + \int_x^y v(\lambda;x,\xi)\,v(\mu;\xi,y)\,d\xi. \quad (9)$$

This relation expresses an *integral addition theorem* for the function $v(\lambda;x,y)$.

The above remarks made for a particular case are true in general, since *by means of the transformation discussed in the preceding section, by which we pass from a power series to a series of powers by composition, all formulae expressing addition theorems in ordinary analysis* (e.g. (7)) *give rise to formulae* (e.g. (9)) *expressing integral addition theorems.* Other examples are the functions by composition that originate from elliptic functions, etc.

Lastly ([1]), we may remind the reader that in the theory of partial differential equations (elementary solution of the linear equation) we also meet integral addition theorems, when from the configuration of a system at an instant t_0, characterised by a solution $\varphi(t_0)$ of the equation, we wish to pass to the configuration at the instant t_0+h+k, characterised by $\varphi(t_0+h+k)$, by passing first from t_0 to t_0+h, and then from t_0+h to t_0+h+k. In this case the total transformation by which we pass from $\varphi(t_0)$ to $\varphi(t_0+h+k)$ is expressed by an integral addition theorem by means of the two partial transformations due to the introduction of the intermediate configuration $\varphi(t_0+h)$.

([1]) Cf. J. HADAMARD, (12).

§ 5. General Theorem on the Solution of Integral Equations.

84. Now let us consider the equation

$$\varphi(z_1, z_2, \ldots z_n) = 0 \qquad (1)$$

or

$$\sum_{0}^{\infty}{}_{i_1 i_2 \ldots i_n} a_{i_1 i_2 \ldots i_n} z_1^{i_1} z_2^{i_2} \ldots z_n^{i_n} = 0 \qquad (1')$$

which is obtained by equating to zero an analytical function φ of the z_r's, regular in the region round the origin, and let us suppose that $\varphi(0, 0, \ldots 0) = 0$ (i.e. $a_{00\ldots 0} = 0$) but $\left(\dfrac{\partial \varphi}{\partial z_n}\right)_{z_1 = z_2 = \cdots = z_n = 0} \neq 0$ (i.e. $a_{00\ldots 1} \neq 0$). With this hypothesis it is known that the equation (1) or (1') defines an implicit function $z_n = \psi(z_1, z_2, \ldots z_{n-1})$, which is regular in the region round $z_1 = z_2 = \cdots = z_{n-1} = 0$, and vanishes at that point. We have therefore

$$\left.\begin{array}{c} z_n = \psi(z_1, z_2, \ldots z_{n-1}) = \sum_{0}^{\infty} b_{i_1 i_2 \ldots i_{n-1}} z_1^{i_1} z_2^{i_2} \ldots z_{n-1}^{i_{n-1}} \\ (b_{00\ldots 0} = 0) \end{array}\right\}. \quad (2)$$

We shall then have, by the remark at the end of the previous section, that the integral (and in general transcendental) equation in the unknown function f_n

$$\varphi(\overset{*}{f_1}, \overset{*}{f_2}, \ldots \overset{*}{f_n}) = \sum_{0}^{\infty} a_{i_1 i_2 \ldots i_n} \overset{*}{f_1}^{i_1} \overset{*}{f_2}^{i_2} \ldots \overset{*}{f_n}^{i_n} = 0 \qquad (3)$$

(in which we suppose $f_1, f_2, \ldots f_{n-1}$ permutable with each other) will have a solution, also permutable with $f_1, f_2, \ldots f_{n-1}$, given by

$$f_n = \psi(\overset{*}{f_1}, \overset{*}{f_2} \ldots \overset{*}{f_{n-1}}) = \sum_{0}^{\infty} b_{i_1 i_2 \ldots i_{n-1}} \overset{*}{f_1}^{i_1} \overset{*}{f_2}^{i_2} \ldots \overset{*}{f_{n-1}}^{i_{n-1}}. \qquad (4)$$

While, however, the series (2) in general converges only in a certain field, the series (4) *always converges* whatever the f's may be (provided they are limited), and will therefore serve to represent the solution in every case[1].

---

[1] J. Pérès has considered the case of the equation (3) and analogous equations when $f_1, f_2, \ldots f_n$ are not permutable with one another. Cf. (19). He has also considered the case in which the series (1') and (2) are not convergent, and gives the conditions which $a_{i_1 i_2 \cdots i_n}$, $b_{i_1 i_2 \cdots i_n}$ must satisfy. Cf. (22), p. 11, and (30).

We thus see that a problem like that of the solution of the integral equation (3), which is more complicated than the solution of the ordinary equation (1) or (1'), has, in a certain sense, a simpler solution, since the series (4) which gives the solution of (3) differs from (2) in being always convergent.

From all the considerations set out up to this point, it follows that, parallel to algebra and the ordinary calculus, which operate on *numbers*, we have succeeded in developing a theory, that of *permutable functions of the first kind*, which from one point of view is of higher degree, since the elements considered in it are not *numbers*, but *functions*, and the operations (composition, functions by composition) are not *functions* applied to numbers, but *functionals* applied to *functions*, but from another is simpler, since all the series which come into consideration are always convergent, and, to use a more expressive phrase, all the singular elements of a function by composition, having its origin in an ordinary analytic function, are removed to infinity.

85. As an example of the solution of a transcendental integral equation, let us consider the equation in the unknown function f

$$g = f + \frac{\overset{*}{f^2}}{2!} + \frac{\overset{*}{f^3}}{3!} + \cdots + \frac{\overset{*}{f^n}}{n!} + \cdots, \qquad (5)$$

which is obtained from the ordinary equation

$$r = z + \frac{z^2}{2!} + \frac{z^3}{3!} + \cdots + \frac{z^n}{n!} + \cdots = e^z - 1, \qquad (6)$$

by substituting the functions g and f for the quantities r and z respectively, and powers by composition for ordinary powers. Solving (6) with respect to z, we have

$$z = \log(1+r) = r - \frac{r^2}{2} + \frac{r^3}{3} - \cdots + (-1)^{n-1}\frac{r^n}{n} + \cdots; \qquad (7)$$

hence the solution f of the transcendental integral equation (5) will be

$$f = g - \frac{\overset{*}{g^2}}{2} + \frac{\overset{*}{g^3}}{3} - \cdots + (-1)^{n-1}\frac{\overset{*}{g^n}}{n} + \cdots, \qquad (8)$$

a series which is always convergent whatever the known function g may be.

86. As an example of the solution of an integral equation of the second degree, let us consider the equation in the unknown f

$$a_1 f + \overset{*}{\varphi_1} \overset{*}{f} + a_2 \overset{*}{f^2} + \overset{*}{\varphi_2} \overset{*}{f^2} = \varphi_0, \qquad (9)$$

or

$$(a_1 \overset{*}{1^0} + \overset{*}{\varphi_1}) \overset{*}{f} + (a_2 \overset{*}{1^0} + \overset{*}{\varphi_2}) \overset{*}{f^2} = \varphi_0,$$

obtained from the ordinary equation

$$(a_1 + u_1) z + (a_2 + u_2) z^2 = u_0 \qquad (9')$$

by substituting f for z, and the functions $\varphi_0, \varphi_1, \varphi_2$ (permutable with each other) for u_0, u_1, u_2.

Since the latter equation has the solution

$$z = \frac{-(a_1 + u_1) + \sqrt{(a_1 + u_1)^2 + 4 u_0 (a_2 + u_2)}}{2(a_2 + u_2)}, \qquad (10)$$

which vanishes for $u_0 = u_1 = u_2 = 0$, the solution of (9) will be obtained from (10) by expanding z as a power series in the u's, and then replacing these powers by the powers by composition of the φ's. It must however be noted that while the ordinary equation (9') has two solutions, the integral equation (9) has only the solution obtained from (10) in the way described above. It can also be shown, in general[1], that the *integral equation of degree m*,

$$(a_m \overset{*}{1^0} + \overset{*}{\varphi_m}) \overset{*}{f^m} + (a_{m-1} \overset{*}{1^0} + \overset{*}{\varphi_{m-1}}) \overset{*}{f^{m-1}} + \cdots + (a_1 \overset{*}{1^0} + \overset{*}{\varphi_1}) \overset{*}{f} = \varphi_0,$$

has always, if a_1 is different from zero, one and only one solution.

§ 6. *Functions Permutable with Unity.*

87. In all the considerations put forward up to now, especially those relative to series and functions by composition, we have always supposed that all the functions $f_1, f_2, \ldots f_n$ that figured

---

[1] Cf. VOLTERRA, (87) of Bibliography I, Chapter IX, § 16.

IV   COMPOSITION AND PERMUTABLE FUNCTIONS   109

simultaneously in the formulae were permutable with each other; this condition being imposed in order that the analogy between *composition* and *ordinary product* might be perfect (i.e. that the commutative property should hold for composition, as well as the associative and distributive properties, and that therefore the ordinary rules of calculation that hold for the algebraic calculus should be applicable without modification (cf. No. 80). A problem of special importance is therefore that of determining the various *aggregates of permutable functions* within each of which, by the considerations set out in the preceding sections, there can be built up a theory of composition; that is to say, a theory dealing with particular functionals (series and functions by composition) defined in the aggregate in question. More precisely, the fundamental problem that we propose to solve is that of *determining all the functions that are permutable with a given one*. The aggregate of all these functions has evidently the character of a *group*, since the composition of any two of them has as its result another function also belonging to the aggregate.

We shall begin by discussing the particular case of the functions $f(x, y)$ that are permutable with a constant, which we can suppose to be unity; then the relation

$$\int_x^y f(x, \xi) d\xi = \int_x^y f(\xi, y) d\xi$$

must hold for $f(x, y)$, and putting $\varphi(x, y)$ for the common value of these two quantities, we shall have, by differentiating,

$$f(x, y) = \frac{\partial \varphi}{\partial y} = -\frac{\partial \varphi}{\partial x},$$

$$\frac{\partial \varphi}{\partial x} + \frac{\partial \varphi}{\partial y} = 0,$$

whence it follows that $\varphi(x, y)$, and therefore also $f(x, y)$, are functions only of the difference $y - x$. Further, it can be shown that all functions of the difference $y - x$ are permutable with one another and with unity. In fact, if we put $\xi - x = y - \eta$, we have

$$\int_x^y f(\xi - x) g(y - \xi) d\xi = \int_x^y g(\eta - x) f(y - \eta) d\eta,$$

i.e.

$$\overset{*}{f}(y - x) \overset{*}{g}(y - x) = \overset{*}{g}(y - x) \overset{*}{f}(y - x).$$

We have therefore the theorem: *All functions that are permutable with unity are also permutable with one another.* They form a particular *group of permutable functions* which VOLTERRA[1] has called the *group of the closed cycle* in connection with its applications to the theory of heredity (see below, Chapter VI, Section IV).

§ 7. *Order of a Function.*

88. Before proceeding to discuss the general case, we shall give some definitions and properties of permutable functions.

If a function $f(x, y)$ can be put in the form

$$f(x, y) = \frac{(y-x)^{\alpha-1}}{\Gamma(\alpha)} \varphi(x, y)$$

(where α is neither 0 nor a negative integer), with the function $\varphi(x, y)$ finite and continuous and $\varphi(x, x) \neq 0$, then we shall say that $f(x, y)$ is *of order* α; $\varphi(x, x)$ will be called the *diagonal* and $\varphi(x, y)$ the *characteristic* of $f(x, y)$. It can then easily be shown[2] that *if two functions are of given orders α and β respectively, then the result of compounding them is of order $\alpha + \beta$, and its diagonal is equal to the product of the diagonals of the two given functions.*

It can also be shown[3] that *if the two functions $f_1(x, y)$, $f_2(x, y)$, of orders α and β respectively, are permutable and their characteristics $\varphi_1(x, y)$, $\varphi_2(x, y)$ are differentiable, then the following relation holds between the diagonals:*

$$\frac{\varphi_1(x, x)^{\frac{1}{\alpha}}}{\varphi_2(x, x)^{\frac{1}{\beta}}} = \text{constant.}$$

§ 8. *Group of the Functions Permutable with a Given Function.*

89. Returning to our fundamental problem of determining all the functions that are permutable with a given function $f(x, y)$ of the first order, we may observe in the first place that we can always suppose that $f(x, y)$ is *in the canonical form*, i.e.

---

[1] Cf. VOLTERRA and PÉRÈS, (44), p. 9.
[2] Cf. VOLTERRA, (42); VOLTERRA and PÉRÈS, (44), p. 11.
[3] Cf. VOLTERRA, (42); VOLTERRA and PÉRÈS, (44), p. 12.

IV COMPOSITION AND PERMUTABLE FUNCTIONS 111

that
$$f(x,x) = 1, \quad \left(\frac{\partial f}{\partial x}\right)_{y=x} = \left(\frac{\partial f}{\partial y}\right)_{y=x} = 0. \tag{1}$$

If it is not, in fact, then by means of the transformation
$$f_1(x_1, y_1) = a(x_1) b(y_1) f(m(x_1), m(y_1)) \tag{2}$$

(in which $a(x_1) b(x_1) = m'(x_1) \neq 0$), we could pass, by a suitable choice of $a(x_1)$, $b(x_1)$, $m(x_1)$, from the original $f(x, y)$ to a function $f_1(x, y)$ in the canonical form(^{1}); and as the transformation (2) by which we pass from f to f_1 does not interfere with composition, i.e. is such that the composition of the two transformed functions f_1 and g_1 has for its result the transformed function of $h = \overset{*}{f}\overset{*}{g}$, i.e.

$$h_1(x_1, y_1) = \overset{*}{f_1}\overset{*}{g_1} = a(x_1) b(y_1) h(m(x_1), m(y_1)),$$

it follows that in order to find all the functions that are permutable with f we need only determine all those that are permutable with f_1 (in the canonical form), and apply to the functions so determined the inverse transformation of (2).

We can therefore limit the problem to finding all the functions $g(x, y)$ that are permutable with a given function $f(x, y)$ *in the canonical form*. Introducing the auxiliary unknown function $\varphi(x, y)$ given by

$$\varphi(x,y) = \int_x^y f(x,\xi) g(\xi,y) d\xi = \int_x^y g(x,\xi) f(\xi,y) d\xi,$$

we shall have
$$\varphi(x,x) = 0, \tag{3}$$

and differentiating, and applying (1),

$$\left.\begin{aligned}\frac{\partial \varphi}{\partial x} &= -g(x,y) + \int_x^y f_1(x,\xi) g(\xi,y) d\xi \\ \frac{\partial \varphi}{\partial y} &= g(x,y) + \int_x^y g(x,\xi) f_2(\xi,y) d\xi\end{aligned}\right\}, \tag{4}$$

where $f_1(x,y) = \dfrac{\partial f(x,y)}{\partial x}$, $f_2(x,y) = \dfrac{\partial f(x,y)}{\partial y}$. Considering (4) as

---
(^{1}) Cf. VOLTERRA and PÉRÈS, (44), p. 38.

two integral equations in the unknown g, and putting

$$F_2 = f_2 - \overset{*}{f_2^2} + \overset{*}{f_2^3} - \overset{*}{f_2^4} + \cdots,$$
$$F_1 = f_1 + \overset{*}{f_1^2} + \overset{*}{f_1^3} + \overset{*}{f_1^4} + \cdots$$

(the solvent kernels of (4)), we shall have

$$g(x,y) = -\frac{\partial \varphi}{\partial x} - \int_x^y F_1(x,\xi) \frac{\partial \varphi(\xi,y)}{\partial \xi} d\xi,$$

$$g(x,y) = \frac{\partial \varphi}{\partial y} - \int_x^y \frac{\partial \varphi(x,\xi)}{\partial \xi} F_2(\xi,y) d\xi;$$

integrating by parts, and putting

$$F_{12} = -\frac{\partial F_2}{\partial x}, \quad F_{21} = -\frac{\partial F_1}{\partial y}$$

(it can be shown(¹) that

$$F_{12} = F_{21} = -(h + \overset{*}{h}\overset{*}{1}\overset{*}{h} + \overset{*}{h}\overset{*}{1}\overset{*}{h}\overset{*}{1}\overset{*}{h} + \cdots) = F,$$

where $h(x,y) = \dfrac{\partial^2 f}{\partial x \partial y}$), we get finally

$$g(x,y) = -\frac{\partial \varphi}{\partial x} - \int_x^y F(x,\xi)\varphi(\xi,y) d\xi,$$

$$g(x,y) = \frac{\partial \varphi}{\partial y} - \int_x^y \varphi(x,\xi) F(\xi,y) d\xi;$$

whence we get for φ the *integro-differential equation* (see also below, Chapter V, § 4)

$$\frac{\partial \varphi}{\partial x} + \frac{\partial \varphi}{\partial y} = \int_x^y [\varphi(x,\xi) F(\xi,y) - F(x,\xi)\varphi(\xi,y)] d\xi. \qquad (5)$$

Calling the right-hand side of this $l(x, y)$, and putting $u = \dfrac{y-x}{2}$, $v = \dfrac{y+x}{2}$, we get from (5) the integral equation

$$\varphi(x,y) = \int_u^v l(\zeta - u, \zeta + u) d\zeta + \vartheta(y-x), \qquad (6)$$

(¹) Cf. J. Pérès, (22).

where ϑ is an arbitrary function which is differentiable and vanishes, like φ, for $x = y$. Putting [1]

$$\vartheta(y - x) = \int_0^{y-x} \lambda(\eta) d\eta,$$

the solution can be obtained by the method of successive approximations by means of the formula

in which
$$\varphi(x, y) = \int_0^{2u} \lambda(\eta) \psi(\eta; x, y) d\eta$$

$$\psi(\eta; x, y) = \sum_{1}^{\infty} \psi_n(\eta; x, y),$$

$$\psi_1 = 1,$$

$$\psi_n = -\int_u^v d\zeta \int_\eta^{2u} d\xi [\psi_{n-1}(\eta; \zeta + u - \xi, \zeta + u) F(\zeta - u, \zeta + u - \xi)$$
$$- \psi_{n-1}(\eta; \zeta - u, \xi + \zeta - u) F(\xi + \zeta - u, \zeta + u)].$$

From this it can easily be deduced [2] that *all the functions $g(x, y)$ that are permutable with $f(x, y)$ are given by the expression*

$$g(x, y) = \lambda(y - x) + \int_0^{y-x} \lambda(\xi) \varphi(\xi; x, y) d\xi, \qquad (7)$$

where $\lambda(y - x)$ is an arbitrary function and $\varphi(\xi; x, y)$ is given by

$$\varphi(\xi; x, y) = \frac{\partial}{\partial y} \psi(\xi; x, y) - \int_{x+\xi}^y d\zeta \psi(\xi; x, \zeta) F(\zeta, y).$$

From (7) it follows in particular that *there exists one and only one function $g(x, y)$ which is permutable with $f(x, y)$ and takes assigned values for $x = 0$*; it is therefore sufficient to determine the corresponding function λ from VOLTERRA's integral equation

$$g(0, y) = \lambda(y) + \int_0^y \lambda(\xi) \varphi(\xi; 0, y) d\xi.$$

The method given here for finding the functions that are permutable with a function $f(x, y)$ of the first order can be

---

[1] Cf. VOLTERRA, (87) of Bibliography I, p. 164.
[2] Cf. VOLTERRA and PÉRÈS, (44), p. 42.

extended to the problem of finding all the functions that are permutable with a function $f(x, y)$ of order higher than the first. The process has been carried out for functions of the second order by VOLTERRA[1], and for those of any integral order whatever, subject to suitable hypotheses on the analytic character of the functions considered, by PÉRÈS[2].

§ 9. *Transformations which maintain the Law of Composition, or Pérès's Transformations.*

90. Formula (7) of the preceding section defines a transformation by which we pass from the group of all the functions $\lambda(y-x)$ that are permutable with unity to the group of all the functions $g(x, y)$ that are permutable with a given function $f(x, y)$ (by means of which the kernel $\varphi(\xi; x, y)$ is constructed). This transformation, which is in substance a *functional* defined in the field of the functions $\lambda(y-x)$ that are permutable with unity, will be denoted shortly by $\Omega(\lambda)$, or if we wish to indicate the kernel $\varphi(\xi; x, y)$ on which it depends, by $\Omega_\varphi(\lambda)$.

It must however be pointed out that there exist more than one of these functionals $\Omega(\lambda)$ which are such that when applied to the functions λ of the closed cycle they generate the group of the functions that are permutable with a given function f. If in fact we apply VOLTERRA's transformation

$$\lambda(\xi) = \mu(\xi) + \int_0^\xi \mu(\eta) k(\eta, \xi) d\eta \tag{1}$$

to the arbitrary function λ, then (7) of the preceding section becomes

$$g(x,y) = \mu(y-x) + \int_0^{y-x} \mu(\xi) \varphi_1(\xi; x, y) d\xi, \tag{2}$$

where μ is arbitrary and

$$\varphi_1(\xi; x, y) = k(\xi, y-x) + \varphi(\xi; x, y)$$
$$+ \int_\xi^{y-x} k(\xi, t) \varphi(t; x, y) dt. \tag{3}$$

---

[1] Cf. VOLTERRA, (40).
[2] Cf. J. PÉRÈS, (21).

IV COMPOSITION AND PERMUTABLE FUNCTIONS 115

We see therefore that (2) is also a transformation $\Omega_{\varphi_1}(\mu)$ which, when applied to the group of the functions $\mu(y-x)$ that are permutable with unity, once again generates the group of the functions that are permutable with $f(x, y)$. The kernel φ_1 which determines this new transformation is given by (3), which, further, gives the most general expression for φ_1, for k an arbitrary function[1].

Among all the transformations $\Omega_\varphi(\lambda)$, depending on an arbitrary function $k(x, y)$ of two variables, we shall now try to determine those (if they exist) that *maintain the property of composition* (PÉRÈS's transformations), i.e. those for which, putting

$$g = \Omega(\lambda), \qquad h = \Omega(\mu),$$

we have

$$\overset{*}{g}\overset{*}{h} = \overset{*}{\Omega}(\lambda)\overset{*}{\Omega}(\mu) = \Omega(\overset{*}{\lambda}\overset{*}{\mu}). \tag{4}$$

The property (4) is easily transformed[2] by means of (1) into the equation

$$\varphi(\tau+\eta; x, y) = \varphi(\tau; x+\eta, y) + \varphi(\eta; x, y-\tau)$$
$$+ \int_{x+\eta}^{y-\tau} \varphi(\eta; x, \xi)\varphi(\tau; \xi, y)\,d\xi, \tag{5}$$

which must be satisfied by the kernel $\varphi(\xi; x, y)$ which is characteristic of Ω.

Now taking $n(x, y)$ arbitrary, and putting

$$m(x, y) = -n(x, y) + \overset{*}{n^2}(x, y) - \overset{*}{n^3}(x, y) + \cdots,$$

(i.e. m is the reciprocal function of n), we get the relation (cf. formula (11) of No. 43)

$$m(x, y) + n(x, y) + \int_x^y m(x, \xi)n(\xi, y)\,d\xi = 0$$

between m and n; i.e.

$$(\overset{*}{1^0} + \overset{*}{m})(\overset{*}{1^0} + \overset{*}{n}) = \overset{*}{1^0} = (\overset{*}{1^0} + \overset{*}{n})(\overset{*}{1^0} + \overset{*}{m}); \tag{6}$$

---

[1] Cf. J. PÉRÈS, (21).
[2] Cf. VOLTERRA and PÉRÈS, (44), p. 58.

and it is easy to verify(¹) that *the most general expression for the kernels φ of the Ω's that maintain the property of composition (an expression depending on the arbitrary function $n(x,y)$ of two variables) is given by*

$$\varphi(\xi; x, y) = n(x + \xi, y) + m(x, y - \xi)$$
$$+ \int_x^{y-\xi} m(x, t) n(t + \xi, y) dt \qquad (7)$$

and the corresponding transformation $\Omega_\varphi(\lambda)$ by

$$g(x, y) = (\overset{*}{1^0} + \overset{*}{n}) \overset{*}{\lambda} (\overset{*}{1^0} + \overset{*}{m}). \qquad (8)$$

In fact, if $h(x, y)$ is any other function whatever given by

$$h(x, y) = \Omega(\mu) = (\overset{*}{1^0} + \overset{*}{n}) \overset{*}{\mu} (\overset{*}{1^0} + \overset{*}{m}),$$

we shall have

$$\overset{*}{\Omega}(\lambda) \overset{*}{\Omega}(\mu) = \overset{*}{g} \overset{*}{h} = (\overset{*}{1^0} + \overset{*}{n}) \overset{*}{\lambda} (\overset{*}{1^0} + \overset{*}{m}) (\overset{*}{1^0} + \overset{*}{n}) \overset{*}{\mu} (\overset{*}{1^0} + \overset{*}{m}),$$

and by (6)

$$\overset{*}{\Omega}(\lambda) \overset{*}{\Omega}(\mu) = (\overset{*}{1^0} + \overset{*}{n}) \overset{*}{\lambda} \overset{*}{\mu} (\overset{*}{1^0} + \overset{*}{m}) = \Omega(\overset{*}{\lambda} \overset{*}{\mu}).$$

The transformation Ω defined by (8), where n and m are connected by (6), is therefore the most general transformation that maintains the property of composition.

In order to show that the group of the functions that are permutable with a given $f(x, y)$ (in the canonical form) can be generated by a Pérès's transformation, i.e. by one of these transformations Ω_φ that maintain the property of composition, we need only show that a kernel φ can be determined in such a way that

$$f(x, y) = \Omega_\varphi(1),$$

since all the g's that are permutable with f will be given by

$$g(x, y) = \Omega_\varphi(\lambda),$$

where λ is any function whatever belonging to the group of the closed cycle.

(¹) Cf. Pérès, (29), and Volterra and Pérès, (44), p. 60.

IV COMPOSITION AND PERMUTABLE FUNCTIONS 117

Such a kernel $\varphi(\xi; x, y)$ is in fact given [1] by the expression

$$\varphi(\xi; x, y) = \sum_{1}^{\infty} \int_0^\xi d\eta_n \ldots \int_0^{\eta_2} d\eta_1 \overset{*}{l}_{\eta_1} \overset{*}{l}_{\eta_2} \ldots \overset{*}{l}_{\eta_n}(x, y - \xi), \quad (9)$$

where $l_\eta(x, y) = l(x+\eta, y+\eta)$ and $l(x, y)$ is determined by the equation

$$f(x, y) = 1 + \overset{*}{1}\overset{*}{l}\overset{*}{1} + \overset{*}{1}\overset{*}{l}\overset{*}{1}\overset{*}{l}\overset{*}{1} + \cdots \quad (10)$$

by means of the formula

$$l(x, y) = -h - \overset{*}{h}\overset{*}{1}\overset{*}{h} - \overset{*}{h}\overset{*}{1}\overset{*}{h}\overset{*}{1}\overset{*}{h} - \cdots, \quad (11)$$

where $h(x, y) = \dfrac{\partial^2 f}{\partial x \partial y}$. We see therefore that the function $l(x, y)$ given by (11), which we have to substitute in (9) in order to get the kernel φ of the transformation, is precisely the function $F(x, y)$ considered in the previous section.

§ 10.

91. Since there does exist a suitable kernel $\varphi(\xi; x, y)$ of a transformation Ω_φ that maintains the property of composition, for which $\Omega_\varphi(1)$ is equal to a given function $f(x, y)$ which is entirely arbitrary, we have, for the group of all the functions $g(x, y)$ that are permutable with $f(x, y)$, given by

$$g(x, y) = \Omega_\varphi(\lambda) = \lambda(y - x) + \int_0^{y-x} \lambda(\xi)\varphi(\xi; x, y) d\xi, \quad (1)$$

the following fundamental theorem: *All the functions that are permutable with a given function f of the first order are permutable with one another.* In fact, if g and h are two such functions obtained by means of the general transformation (1) from two other functions λ and μ respectively, belonging to the group of the closed cycle, then

$$g = \Omega(\lambda), \quad h = \Omega(\mu),$$

$$\overset{*}{g}\overset{*}{h} = \overset{*}{\Omega}(\lambda)\overset{*}{\Omega}(\mu);$$

and since Ω is a PÉRÈS's transformation (maintaining the property of composition), it follows (taking into account the fact

---

[1] Cf. PÉRÈS, (26).

118   THEORY OF FUNCTIONALS   CHAP.

that λ and μ are permutable with unity, and therefore with each other) that

$$\overset{*}{g}\overset{*}{h} = \overset{*}{\Omega}(\lambda)\overset{*}{\Omega}(\mu) = \Omega(\overset{*}{\lambda}\overset{*}{\mu}) = \Omega(\overset{*}{\mu}\overset{*}{\lambda}) = \overset{*}{\Omega}(\mu)\overset{*}{\Omega}(\lambda) = \overset{*}{h}\overset{*}{g}.$$

The proof given here of this theorem is that of PÉRÈS, but it had already been enunciated by VOLTERRA and proved subsequently by VESSIOT([1]) and by VOLTERRA([2]).

§ 11. *Powers by Composition with a Fractional Exponent.*

92. We have already (No. 81) defined powers by composition with a positive integral exponent, and the power with exponent 0. If, given a function $h(x, y)$, there exists another function $g(x, y)$ such that $\overset{*}{g}{}^n = h$ (n a positive integer), we shall say that *g is a power of h whose exponent is* $\dfrac{1}{n}$, and shall write

$$g(x, y) = \overset{*}{h}{}^{\frac{1}{n}}.$$

We shall then define the power of h with exponent $\dfrac{p}{n}$ (a positive fraction) by means of the equation

$$\overset{*}{h}{}^{\frac{p}{n}} = \overset{*}{g}{}^p \qquad (p \text{ an integer}).$$

Hence to define the fractional powers by composition of a function h, we shall have to solve binomial equations of the type

$$\overset{*}{g}{}^n(x, y) = h(x, y) \qquad (1)$$

in the unknown $g(x, y)$. If the function h is of order β (positive), g, if its order is determinate, must be of order $\dfrac{\beta}{n}$ (cf. No. 88). We propose to show that *we can solve the equation* (1) *whenever it is possible to determine a function* $\vartheta(x, y)$ *of order* $\dfrac{\beta}{n}$ *which is permutable with* $h(x, y)$.

In this case, in fact, $\overset{*}{\vartheta}{}^n$ will be of order β and permutable with

---
([1]) Cf. E. VESSIOT, (35), (36).
([2]) Cf. VOLTERRA, (42), and (44), p. 66.

IV COMPOSITION AND PERMUTABLE FUNCTIONS 119

h; hence (No. 88) the ratio of the diagonals of h and $\overset{*}{\vartheta}{}^n$ will be a constant c, and the function $h - c\overset{*}{\vartheta}{}^n$ will be of order higher than β. But then the integral equation

$$h - c\overset{*}{\vartheta}{}^n = c\overset{*}{\varphi}\overset{*}{\vartheta}{}^n, \qquad (2)$$

subject to suitable hypotheses as to differentiability (cf. No. 48), will have a solution φ permutable with ϑ, and therefore

$$h = c\overset{*}{\vartheta}{}^n(\overset{*}{1}{}^0 + \overset{*}{\varphi}).$$

The function φ being now determined by (2), in order to define $h^{\frac{1}{n}} = g$ we have only to put

$$g = h^{\frac{1}{n}} = \sqrt[n]{c}\,\overset{*}{\vartheta}(\overset{*}{1}{}^0 + \overset{*}{\varphi})^{\frac{1}{n}}$$

$$= \sqrt[n]{c}\left(\vartheta + \frac{1}{n}\overset{*}{\vartheta}\overset{*}{\varphi} + \frac{\frac{1}{n}\left(\frac{1}{n}-1\right)}{2!}\overset{*}{\vartheta}\overset{*}{\varphi}{}^2 + \cdots\right).$$

If in particular h is permutable with a function $f(x, y)$ of the first order (which we can always suppose in the canonical form; cf. No. 89), then having determined (as in the previous section) a transformation $\Omega_\varphi(\lambda)$ which generates all the functions that are permutable with f, we need only take for ϑ the following function:

$$\vartheta(x, y) = (y - x)^{\frac{\beta}{n}-1} + \int_0^{y-x} \xi^{\frac{\beta}{n}-1} \varphi(\xi; x, y)\, d\xi.$$

93. Powers by composition with any index whatever. If h is of given order α, i.e. if

$$h(x, y) = \frac{(y-x)^{\alpha-1}}{\Gamma(\alpha)}\bar{h}(x, y), \quad \bar{h}(x, x) \neq 0,$$

and if the powers with a fractional exponent $\overset{*}{h}{}^{\frac{p}{n}}$ and of order $\left(\frac{p}{n}\alpha\right)$ are defined by the formula

$$\overset{*}{h}{}^{\frac{p}{n}} = \frac{(y-x)^{\frac{p}{n}\alpha-1}}{\Gamma\left(\frac{p}{n}\alpha\right)} l\left(x, y; \frac{p}{n}\right)$$

where $l\left(x, x; \dfrac{p}{n}\right) = [h(x, x)]^{\frac{p}{n}}$, we can in many cases pass on to a definition of powers with an irrational exponent. In fact, if $l\left(x, y; \dfrac{p}{n}\right)$ tends uniformly to $l(x, y; \gamma)$ as $\dfrac{p}{n}$ tends to a rational number γ, and if it tends uniformly to a definite finite limit, which we shall denote by $l(x, y; z)$, as $\dfrac{p}{n}$ tends to an irrational number z, then by definition we shall put

$$\overset{*}{h}{}^{z} = \frac{(y-x)^{\alpha z - 1}}{\Gamma(\alpha z)} l(x, y; z).$$

The whole algebraic calculus of ordinary powers can be extended to these powers by composition with a real positive exponent; among other results, we get the formulae

$$\overset{*}{h}{}^{z} \overset{*}{h}{}^{z_1} = \overset{*}{h}{}^{z+z_1}, \quad (\overset{*}{h}{}^{z})^{z_1} = \overset{*}{h}{}^{zz_1}.$$

If in particular we take $h = 1$, we have

$$\overset{*}{1}{}^{\frac{p}{q}} = \frac{(y-x)^{\frac{p}{q}-1}}{\Gamma\left(\dfrac{p}{q}\right)},$$

and therefore, for any positive quantity z,

$$\overset{*}{1}{}^{z} = \frac{(y-x)^{z-1}}{\Gamma(z)}$$

(an integral function of z). If instead h is a function belonging to the group of the closed cycle of the first order, differentiable and with its diagonal $= 1$, we have

$$h = 1 + \overset{*}{1} \overset{*}{h}{}' = \overset{*}{1}(\overset{*}{1}{}^{0} + \overset{*}{h}{}')$$

where h' denotes the derivative of $h(y-x)$ with respect to y, and therefore

$$\overset{*}{h}{}^{z} = \overset{*}{1}{}^{z}(\overset{*}{1}{}^{0} + \overset{*}{h}{}')^{z} = \overset{*}{1}{}^{z} + z \overset{*}{1}{}^{z} \overset{*}{h}{}' + \frac{z(z-1)}{2!} \overset{*}{1}{}^{z} \overset{*}{h}{}'^{2} + \cdots,$$

and $\overset{*}{h}{}^{z}$ too will be an *integral function of z*.

IV COMPOSITION AND PERMUTABLE FUNCTIONS 121

If h is, in general, a function of the first order, which we may suppose in the canonical form, we can define $\overset{*}{h}{}^z$ by observing that if Ω_φ is a Pérès's transformation (maintaining the property of composition) for which

$$h(x,y) = \Omega_\varphi(1)$$

(cf. No. 91), then the power $\overset{*}{h}{}^z$ will be given by

$$\overset{*}{h}{}^z = \overset{*}{\Omega}{}^z_\varphi(1) = \Omega(\overset{*}{1}{}^z) = \frac{(y-x)^{z-1}}{\Gamma(z)} + \int_0^{y-x} \frac{\xi^{z-1}}{\Gamma(z)} \varphi(\xi; x,y) d\xi.$$

In this very general case also $\overset{*}{h}{}^z$ is therefore an integral function of z.
This theorem was first enunciated by VOLTERRA [1] and was subsequently proved by PÉRÈS [2].

§ 12. Asymptotic Methods.

94. Another method, which we may call asymptotic, of calculating the mth powers by composition of a function f consists in putting

$$[\overset{*}{f}{}^0 + z(\overset{*}{f} - \overset{*}{f}{}^0)]^m = [(1-z)\overset{*}{f}{}^0 + z\overset{*}{f}]^m$$
$$= (1-z)^m \left[\overset{*}{f}{}^0 + \binom{m}{1} \frac{z}{1-z} f + \binom{m}{2} \left(\frac{z}{1-z}\right)^2 \overset{*}{f}{}^2 + \cdots \right],$$

and then passing to the limit for $z=1$. In fact, when this limit exists, we shall have

$$\overset{*}{f}{}^m = \lim_{z=1} [\overset{*}{f}{}^0 + z(\overset{*}{f} - \overset{*}{f}{}^0)]^m.$$

In particular, if f is of order n (No. 88), we can by this method calculate $\overset{*}{f}{}^{\frac{1}{n}}$, which will be of the first order. If then we wish to determine all the functions that are permutable with a function f of any integral order n whatever (a problem solved by PÉRÈS for the single case of f an analytic function), we have

---

[1] Conférence au collège de France, 1919.
[2] Cf. J. PÉRÈS, (28).

only to determine the group of the functions that are permutable with $\overset{*}{f}{}^{\frac{1}{n}}$ (a function of the first order), and these will be also permutable with $f = (\overset{*}{f}{}^{\frac{1}{n}})^n$ (¹).

§ 13. Fractions by Composition and Powers by Composition with a Negative Exponent.

95. Following out the analogy with the methods of ordinary arithmetic, as we have done for powers, if we have two given functions f and g belonging to a group of permutable functions, then the symbol $\dfrac{\overset{*}{g}}{\underset{*}{f}}$ will be, by definition, the *fraction by composition* belonging to the group, with the numerator g and the denominator f. Two fractions by composition $\dfrac{\overset{*}{g_1}}{\underset{*}{f_1}}$ and $\dfrac{\overset{*}{g_2}}{\underset{*}{f_2}}$ will be said to be *equal* if $\overset{*}{g_1}\overset{*}{f_2} = \overset{*}{g_2}\overset{*}{f_1}$; and it is easy to verify that with this definition the fundamental property of equality is preserved, viz. *two fractions by composition which are equal to a third are equal to each other* (²). It can also be proved at once that

$$\frac{\overset{*}{g}}{\underset{*}{f}} = \frac{\overset{**}{gh}}{\underset{**}{fh}}. \tag{1}$$

Advantage can be taken of this property to simplify a fraction by composition. For example, if g is of higher order than f, there will be a true (not symbolical) function h such that $g = \overset{**}{fh}$, and therefore

$$\frac{\overset{*}{g}}{\underset{*}{f}} = \frac{\overset{**}{fh}}{\underset{**}{fh^0}} = \frac{\overset{*}{h}}{\underset{*}{h^0}} = h.$$

In this case (the only one), therefore, the fraction by composition $\dfrac{\overset{*}{g}}{\underset{*}{f}}$ is not merely a symbol, but is **a true function** h

---

(¹) Cf. VOLTERRA, (45).
(²) Cf. VOLTERRA, (42), and VOLTERRA and PÉRÈS, (44), p. 89.

IV COMPOSITION AND PERMUTABLE FUNCTIONS 123

which when compounded with the denominator f gives as result the numerator g.

It can also be seen at once, in virtue of the property (1), that several fractions by composition can be reduced to a common denominator; hence we have an obvious definition of the *sum* (or *difference*) of two or more fractions by composition. The composition of two or more fractions will be the result of compounding all the numerators and all the denominators.

96. The foregoing definitions and rules have as a consequence the formula

$$\frac{\overset{*}{g}}{\overset{*}{f}} \overset{*}{f} = g,$$

i.e. the result of compounding a fraction with its denominator is its numerator. If then we define the symbol $\overset{*}{f}^{-1}$ by means of the formula $\overset{*}{f}^{-1} \overset{*}{f} = \overset{*}{f}{}^0$, we can observe that $\overset{**}{g f^{-1}}$ also has the property that $(\overset{**}{g f^{-1}}) \overset{*}{f} = g$. It is therefore natural to write

$$\frac{\overset{*}{g}}{\overset{*}{f}} = \overset{**}{g f^{-1}} = \overset{**}{f^{-1} g}.$$

Having thus defined the power $\overset{*}{f}^{-1}$ with exponent -1, in order to define $\overset{*}{f}^{-m}$, where m is any positive quantity (cf. No. 95), we shall put

$$\overset{*}{f}^{-m} = (\overset{*}{f}{}^m)^{-1}.$$

We may observe (¹) that these powers by composition with a negative index, which up to now have a purely symbolic significance, can in many cases be replaced by *true functions*. For this we need only define *functions of order* α *(where* α *may be negative, but not* 0 *or a negative integer)* as those functions f which can be put in the form

$$f(x, y) = \frac{(y-x)^{\alpha-1}}{\Gamma(\alpha)} \varphi(x, y)$$

in which always $\varphi(x, x) \neq 0$. Functions of order α (α not 0 or a negative integer) will be called functions of *regular order*. The

---
(¹) Cf. J. Pérès, (23).

composition of two functions f and g of regular orders α und β will then be defined, if $\alpha + \beta$ is also regular, by means of the usual formula

$$\overset{**}{fg} = \int_x^y f(x,\xi)\,g(\xi,y)\,d\xi\,,$$

it being agreed, if the integral is infinite, to take its finite part([1])

Furthermore, given any relation whatever containing symbolic expressions (fractions, powers by composition with zero or negative exponent), all that is necessary to transform the original relation into a relation between ordinary functions is to compound all its terms with a function of sufficiently high positive order.

§ 14. *Logarithms by Composition.*

97. We shall call the sequence

$$\ldots \overset{*}{f}{}^{-2},\ \overset{*}{f}{}^{-1},\ \overset{*}{f}{}^{0},\ f,\ \overset{*}{f}{}^{2} \ldots$$

a *progression by composition*.

The exponents will be called the *logarithms by composition* of the various powers, f will be the *base*, and we shall write

$$\log_f \overset{*}{f}{}^m = m$$

or alternatively, for the sake of homogeneity,

$$\log_f \overset{*}{f}{}^m = \log_f f^m \overset{*}{f}{}^{0} = m \overset{*}{f}{}^{0},$$

even if m is not an integer.

Now let us consider the function $e\overset{*}{f}{}^{0} = \overset{*}{e}$, of order zero; we have $\overset{*}{e^z} = \overset{*}{e^z} \overset{*}{f}{}^{0} = e^z \overset{*}{f}{}^{0}$, and therefore

$$\frac{d}{dz}\overset{*}{e^z} = e^z \overset{*}{f}{}^{0} = \overset{*}{e^z}.$$

We shall take $\overset{*}{e}$ as base of a system of logarithms by composition, which will therefore be called *napierian*.

([1]) For the definition of the *finite part* of an improper integral, and for its properties, cf. HADAMARD, (11).

IV COMPOSITION AND PERMUTABLE FUNCTIONS

As we have
$$\overset{*}{e}{}^z = \overset{*}{e}{}^z \overset{*}{f}{}^0 = (z\overset{*}{f}{}^0)^0 + z\overset{*}{f}{}^0 + \frac{(z\overset{*}{f}{}^0)^2}{2!} + \cdots + \frac{(z\overset{*}{f}{}^0)^n}{n!} + \cdots,$$
we shall by definition put
$$\overset{*}{e}{}^\varphi = \overset{*}{\varphi}{}^0 + \varphi + \frac{\overset{*}{\varphi}{}^2}{2!} + \cdots + \frac{\overset{*}{\varphi}{}^n}{n!} + \cdots, \tag{1}$$
when φ is any function whatever; and if
$$\overset{*}{e}{}^\varphi = \psi, \tag{2}$$
we shall say, also by definition, that φ is the logarithm by composition, to the base $\overset{*}{e}$, of ψ, and we shall write
$$\varphi = \overset{*}{\log}_e \psi, \quad \text{or} \quad \varphi = \overset{*}{l}\psi. \tag{3}$$

We have already considered the series (1) in No. 85; formula (8) of No. 83 is a particular case of
$$\overset{*}{e}{}^\varphi \overset{*}{e}{}^\chi = \overset{*}{e}{}^{\varphi+\chi}.$$
Similarly
$$\overset{*}{e}{}^{2\pi i \overset{*}{\varphi}{}^0} = \overset{*}{\varphi}{}^0, \quad \overset{*}{e}{}^{\varphi + 2n\pi i \overset{*}{\varphi}{}^0} = \overset{*}{e}{}^\varphi \quad (n \text{ an integer}).$$

98. We shall now try to determine the napierian logarithm by composition of a function ψ. If in (2) φ is a function of positive order, we have
$$\overset{*}{e}{}^\varphi = \overset{*}{1}{}^0 + \Theta,$$
where
$$\Theta = \varphi + \frac{\overset{*}{\varphi}{}^2}{2!} + \cdots + \frac{\overset{*}{\varphi}{}^n}{n!} + \cdots. \tag{4}$$

If the function ψ whose logarithm we wish to determine is of the form
$$\psi = \overset{*}{1}{}^0 + h, \tag{5}$$
then to determine $\overset{*}{l}\psi = \varphi$ we have to solve the integral equation
$$h = \varphi + \frac{\overset{*}{\varphi}{}^2}{2!} + \cdots + \frac{\overset{*}{\varphi}{}^n}{n!} + \cdots,$$

and we shall therefore (cf. No. 85) have

$$\varphi = \overset{*}{l}\psi = h - \frac{\overset{*}{h}{}^2}{2} + \frac{\overset{*}{h}{}^3}{3} - \cdots + (-1)^{n-1}\frac{\overset{*}{h}{}^n}{n} + \cdots \quad (5')$$

If ψ is not of the form (5), it follows from (3) that

$$\overset{*}{\psi}{}^z = \overset{*}{e}{}^{z\varphi},$$

and differentiating with respect to z,

$$\frac{d}{dz}\overset{*}{\psi}{}^z = \overset{*}{e}{}^{z\varphi}\overset{*}{\varphi} = \overset{*}{\psi}{}^z\overset{*}{\varphi},$$

from which finally we get

$$\varphi = \overset{*}{l}\psi = \overset{*}{\psi}{}^{-z}\frac{d\overset{*}{\psi}{}^z}{dz}. \quad (6)$$

We thus get $\overset{*}{l}\psi$ in the form of a *fraction by composition*, which can easily be shown to be independent of z([1]) if $\psi(x, y)$ is of order $\alpha > 0$ and

$$\overset{*}{\psi}{}^z = \frac{(y-x)^{\alpha z - 1}}{\Gamma(\alpha z)}\overline{\psi}(x, y; z)$$

is an analytic function of z.

99. As an example let us calculate $\overset{*}{l}1$. Since

$$\overset{*}{1}{}^z = \frac{(y-x)^{z-1}}{\Gamma(z)}$$

(No. 93), we shall have([2])

$$\frac{d\overset{*}{1}{}^z}{dz} = \frac{(y-x)^{z-1}}{\Gamma(z)}\log(y-x) - \frac{(y-x)^{z-1}}{\Gamma^2(z)}\Gamma'(z) = \vartheta(x, y; z),$$

and

$$\overset{*}{l}1 = \overset{*}{\vartheta}\overset{*}{1}{}^{-z}.$$

([1]) Cf. VOLTERRA and PÉRÈS, (44), p. 121.

([2]) To avoid ambiguity we write log A to represent the napierian arithmetical logarithm of the number A.

IV COMPOSITION AND PERMUTABLE FUNCTIONS 127

Since $\overset{*}{l}1$ is independent of z, we can put $z=1$ in this relation, and we thus obtain

$$\vartheta(x,y;1) = \log(y-x) - \Gamma'(1) = \log(y-x) + C$$

($C = 0.57721\ldots$, EULER's constant), and therefore

$$\overset{*}{l}1 = \frac{\overline{\log(y-x) + C}}{\overset{*}{1}}. \tag{7}$$

100. The following properties hold for logarithms by composition:

$$\overset{*}{l}(\overset{*}{\psi}\overset{*}{\vartheta}) = \overset{*}{l}\psi + \overset{*}{l}\vartheta, \tag{a}$$

$$\overset{*}{l}\frac{\overset{*}{\psi}}{\overset{*}{\vartheta}} = \overset{*}{l}\psi - \overset{*}{l}\vartheta, \tag{b}$$

$$\overset{*}{l}(\psi^m) = m\overset{*}{l}\psi, \tag{c}$$

$$\overset{*}{l}(c\,\vartheta) = \log c\,\overset{*}{\vartheta}{}^0 + \overset{*}{l}\vartheta; \tag{d}$$

or, in another form,

$$\overset{*}{e}{}^{\varphi+\chi} = \overset{*}{e}{}^\varphi\,\overset{*}{e}{}^\chi, \tag{a'}$$

$$\overset{*}{e}{}^{\varphi-\chi} = \frac{\overset{*}{e}{}^\varphi}{\overset{*}{e}{}^\chi}, \tag{b'}$$

$$(\overset{*}{e}{}^\varphi)^m = \overset{*}{e}{}^{m\varphi}, \tag{c'}$$

$$\overset{*}{e}{}^c\overset{*}{\chi}{}^0{}^{+\chi} = e^c\,\overset{*}{e}{}^\chi. \tag{d'}$$

The formulae (c) and (c'), which have been proved for m real, can be extended by definition to the case of m any complex quantity whatever; we shall therefore have

$$\overset{*}{e}{}^{m\varphi} = (m\overset{*}{\varphi})^0 + m\varphi + \frac{(m\overset{*}{\varphi})^2}{2!} + \cdots.$$

Hence, if $\psi = \overset{*}{e}{}^\varphi$, $\overset{*}{\psi}{}^m$ will be defined, for m any complex quantity whatever, by

$$\overset{*}{\psi}{}^m = \overset{*}{e}{}^{m\overset{*}{l}\psi}.$$

Further, if φ and ϑ are permutable, we shall also put by definition
$$(\overset{*}{e}{}^{\varphi})^{\vartheta} = \overset{**}{e}{}^{\overset{*}{\varphi}\vartheta},$$
and, if $\overset{*}{e}{}^{\varphi} = \psi$,
$$\overset{*}{\psi}{}^{\vartheta} = \overset{**}{e}{}^{\overset{*}{\vartheta}\overset{*}{l}\psi} = \chi. \tag{8}$$

Extending again the definition of a logarithm, we shall in this case, also by definition, put
$$\vartheta = \overset{*}{\log}_{\psi}\chi,$$
and we shall say that ϑ is the *logarithm by composition of χ to the base ψ*.

From (8) it follows that
$$\overset{*}{\log}_{\psi}\chi = \frac{\overset{*}{l}\chi}{\overset{*}{l}\psi}. \tag{9}$$

101. If we wish to determine the napierian logarithm of a function $f(y-x)$ belonging to the group of the closed cycle, differentiable and of the first order ($f(0)=1$), then since
$$f = \overset{*}{1}(\overset{*}{1}{}^{0} + \overset{*}{f}'),$$
we shall have
$$\overset{*}{l}f = \overset{*}{l}1 + \overset{*}{l}(\overset{*}{1}{}^{0} + \overset{*}{f}'),$$
and by (5')
$$\overset{*}{l}f = \overset{*}{l}1 + f' - \frac{\overset{*}{f}'^{2}}{2} + \frac{\overset{*}{f}'^{3}}{3} - \cdots$$
$$= \overset{*}{l}1 + \varphi, \tag{10}$$
where $\overset{*}{l}1$ is a symbolical function and φ is a true (not symbolical) function.

If we wish to find the logarithm of f to the base 1, we have, by (9),
$$\overset{*}{\log}_{1}f = \frac{\overset{*}{l}f}{\overset{*}{l}1} = \frac{\overset{**}{1}\overset{*}{l}f}{\overset{**}{1}\overset{*}{l}1} = \frac{\overset{**}{1}\overset{*}{l}1 + \overset{**}{1}\varphi}{\overset{*}{\log}(y-x) + C},$$
and we therefore have to solve a VOLTERRA's integral equation of the first kind with kernel $\log(y-x) + C$, an equation that we have already studied in No. 51 above.

IV   COMPOSITION AND PERMUTABLE FUNCTIONS   129

More generally, given two permutable functions ψ and χ, if

$$\overset{*}{\psi}{}^z = (y-x)^{\alpha z - 1} h(x, y; z),$$

$$\frac{d\overset{*}{\psi}{}^z}{dz} = \alpha (y-x)^{\alpha z - 1} h(x, y; z) \log(y-x) + (y-x)^{\alpha z - 1} h'_z(x, y; z),$$

then to determine $\log_\psi \chi = \dfrac{\overset{*}{l}\chi}{\overset{*}{l}\psi} = \dfrac{\overset{*}{\psi}{}^z \overset{*}{l}\chi}{\overset{*}{\psi}{}^z \overset{*}{l}\psi}$ we should get a Volterra's integral equation of the first kind with a *logarithmic kernel*, given by([1])

$$\overset{*}{\psi}{}^z \overset{*}{l}\psi = \frac{d\overset{*}{\psi}{}^z}{dz}.$$

§ 15. *Functions by Composition.*

102. We have already, in No. 82, considered some particular functionals depending on a function of two variables f and on two parameters x and y, of the type

$$g(x, y) = \mathrm{F}[f; x, y] = a_1 f + a_2 \overset{*}{f}{}^2 + a_3 \overset{*}{f}{}^3 + \cdots + a_n \overset{*}{f}{}^n + \cdots, \quad (1)$$

functionals which we have called *series by composition* and *functions by composition*. We may now observe that they all have the two following fundamental properties:

(i) The functional $\mathrm{F}[f; x, y] = g(x, y)$, considered as a function of the two parameters, is a function permutable with f.

(ii) *If we take any other function $\varphi(x, y)$ which is permutable with f, then the limit*

$$\lim_{\varepsilon = 0} \frac{\overset{*}{\mathrm{F}}[f + \varepsilon \varphi] - \overset{*}{\mathrm{F}}[f]}{\varepsilon \overset{*}{\varphi}}$$

exists and is independent of φ.

This limit will be called the *derivative by composition* of F with respect to f, and will be denoted by $\dfrac{d\overset{*}{\mathrm{F}}}{d\overset{*}{f}}$; in the case of (1) we obtain easily

---
([1]) Cf. Volterra, (42); Volterra and Pérès, (44), p. 127.

$$\frac{d\overset{*}{F}}{d\overset{*}{f}} = \lim_{\varepsilon=0} \frac{\overset{*}{F}[f+\varepsilon\varphi] - \overset{*}{F}[f]}{\varepsilon\varphi}$$

$$= a_1 \overset{*}{f^0} + 2a_2 \overset{*}{f} + 3a_3 \overset{*}{f^2} + \cdots + na_n \overset{*}{f^{n-1}} + \cdots. \quad (2)$$

Passing now from these particular examples to a general concept, we shall say that *a functional* $F[f; x, y]$, *depending on a function f of two variables and on two parameters x and y, is a function by composition (of the first kind) of f if it has the two fundamental properties enunciated above*.

We shall thus take these two properties, which hold in all the particular cases considered up to this point, as the basis of the definition of a *function by composition*; and we may observe that the second is analogous to the condition of monogeneity $\left(\lim_{\varepsilon=0} \frac{f(z+\varepsilon z') - f(z)}{\varepsilon z'} \right.$ independent of the complex number $z' \Big)$ which serves as basis for the definition of a function $f(z)$ of a complex variable.

If in particular, to calculate the derivative $\dfrac{d\overset{*}{F}}{d\overset{*}{f}}$, we give f the increment $\varepsilon \overset{*}{f^0}$, we shall have

$$\frac{d\overset{*}{F}}{d\overset{*}{f}} = \lim_{\varepsilon=0} \frac{F[f+\varepsilon \overset{*}{f^0}] - F[f]}{\varepsilon} = \frac{d}{dz}(F[f + z\overset{*}{f^0}])_{z=0},$$

by means of which the calculation of the derivative by composition $\dfrac{d\overset{*}{F}}{d\overset{*}{f}}$ is reduced to the calculation of an ordinary derivative.

It is possible to show (¹) that *the derivative by composition* $\dfrac{d\overset{*}{F}}{d\overset{*}{f}}$ *is itself a new function by composition of f*.

Vice versa, given a function by composition $\overset{*}{\Phi}[f]$, we can set ourselves the problem of determining those functions by

---

(¹) Cf. VOLTERRA, (42); VOLTERRA and PÉRÈS, (44), p. 152.

IV COMPOSITION AND PERMUTABLE FUNCTIONS

composition $F[\overset{*}{f}]$ which have $\overset{*}{\Phi}[f]$ for their derivative; it will be natural to denote these functions by

$$\int \overset{*}{\Phi} d\overset{*}{f} = F[\overset{*}{f}]$$

For this we must first define the *definite integral by composition* in the following way. If $f(x, y; s)$ is a simple infinity of functions (depending on the parameter s), belonging to a group of permutable functions, the result of compounding $\Phi[f(x, y; s)]$ with $\dfrac{\partial f(x, y; s)}{\partial s}$ will also be a function of the parameter s; and if we put

$$f(x, y; a) = f_1(x, y), \qquad f(x, y; b) = f_2(x, y),$$

we shall by definition put

$$\int_{f_1}^{f_2} \overset{*}{\Phi} d\overset{*}{f} = \int_a^b \overset{*}{\Phi}[f(x, y; s)] \frac{\partial \overset{*}{f}(x, y; s)}{\partial s} ds. \qquad (3)$$

This definition and this notation are acceptable since we can prove[1] a theorem analogous to Cauchy's theorem for functions of a complex variable, viz. that the foregoing expression depends only on the two functions f_1 and f_2 which are the limits of the integral, and not on the path traversed to go from f_1 to f_2, i.e. does not depend on the intermediate functions

$$f(x, y; s), \quad \text{with} \quad a < s < b.$$

More precisely, it can be proved that if $f(x, y; u, v)$ is an aggregate of permutable functions $f(x, y)$, and if both f and $\overset{*}{\Phi}[f]$ are regular functions (continuous and differentiable) of u and v when the point whose coordinates are u and v varies within an area σ with boundary s, then the integral by composition

$$\int_s \overset{*}{\Phi}[f] d\overset{*}{f}$$

is identically zero.

It can then easily be seen[2] that if in the integral by composition (3) we suppose the upper limit f_2 variable, the integral

---

[1] Cf. VOLTERRA, (42); VOLTERRA and PÉRÈS, (44), p. 159.
[2] Cf. VOLTERRA, (42); VOLTERRA and PÉRÈS, (44), p. 161.

itself will be a function by composition $\overset{*}{F}[f_2]$ of the limit f_2, and also

$$\frac{d\overset{*}{F}}{d\overset{*}{f_2}} = \overset{*}{\varPhi}[f_2].$$

§ 16. Series by Composition of the Second Kind.

103. We have already seen in No. 82 that if

$$\varphi(z) = a_1 z + a_2 z^2 + \cdots + a_n z^n + \cdots \qquad (1)$$

is a power series with radius of convergence not 0, then the series by composition of the first kind,

$$a_1 z f + a_2 z^2 \overset{*}{f^2} + \cdots + a_n z^n \overset{*}{f^n} + \cdots \qquad (2)$$

will always be convergent for any z and f whatever, i.e. it is an *integral function of z*.

If now we consider the series

$$a_1 z f + a_2 z^2 \overset{**}{f^2} + a_3 z^3 \overset{**}{f^3} + \cdots + a_n z^n \overset{**}{f^n} + \cdots, \qquad (3)$$

which we shall call a *series by composition of the second kind*, this series is a functional of f, but it is no longer generally an integral function of z.

We have for the series (3) the following fundamental theorem:

If the function $\varphi(z)$ given by (1) is an integral function, then the series (3) is also an integral function of z; while if $\varphi(z)$ is the ratio of two integral functions (a meromorphic function of z), then the series (3) is also the ratio of two integral functions of z (a meromorphic function of z).

VOLTERRA has given a direct proof of this theorem using the method of passing from the finite to the infinite (cf. No. 3)[1]. Another proof of this theorem has been given by LEBESGUE[2].

If then we know the singular points z_i (the characteristic values; cf. No. 45) of

$$S(x, y; z) = z f + z^2 \overset{**}{f^2} + z^3 \overset{**}{f^3} + \cdots + z^n \overset{**}{f^n} + \cdots \qquad (4)$$

---

[1] Cf. VOLTERRA, (87) of Bibliography I, p. 192; cf. also Chapter V, § 7.
[2] Cf. LEBESGUE, (15).

(the reciprocal or solvent kernel of the second kind of $-zf$) and the singular points z_j of $\varphi(z)$, it follows from a well-known theorem due to HADAMARD[1] that *the singular points z of the power series* (3), *which has for coefficients the products of the corresponding coefficients of* (1) *and* (4), *will all be included among the points* z_i, z_j; and as both z_i and z_j are independent of x and y, we see thus that the positions of the singular points z of (3), which is also a function of x and y, *also do not depend either on x or on y* (i.e. they are fixed singular points).

104. For these series by composition of the second kind (3) certain properties also hold which are wholly analogous to those found for series by composition of the first kind. Thus, for example, for the integral function $v(z; x, y)$ of No. 83, given by

$$v(z;x,y) = zf + \frac{z^2}{2!}\overset{*}{f^2} + \cdots + \frac{z^n}{n!}\overset{*}{f^n} + \cdots = \overset{*}{\mathrm{F}}[f],$$

the integral addition theorem ((9) of No. 83)

$$v(z_1+z_2;x,y) = v(z_1;x,y) + v(z_2;x,y) + \int_x^y v(z_1;x,\xi)\,v(z_2;\xi,y)\,d\xi$$

holds, and also, if f_1, f_2 are permutable of the first kind,

$$\overset{*}{\mathrm{F}}[f_1+f_2] = \overset{*}{\mathrm{F}}[f_1] + \overset{*}{\mathrm{F}}[f_2] + \overset{*}{\mathrm{F}}[f_1]\overset{*}{\mathrm{F}}[f_2];$$

corresponding to this we get for the integral function given by

$$u(z;x,y) = zf + \frac{z^2}{2!}\overset{**}{f^2} + \cdots + \frac{z^n}{n!}\overset{**}{f^n} + \cdots = \overset{**}{\varPhi}[f]$$

(a series by composition of the second kind) the *integral addition theorem of the second kind*

$$u(z_1+z_2;x,y) = u(z_1;x,y) + u(z_2;x,y)$$
$$+ \int_a^b u(z_1;x,\xi)\,u(z_2;\xi,y)\,d\xi, \quad (5)$$

and also

$$\overset{**}{\varPhi}[f_1+f_2] = \overset{**}{\varPhi}[f_1] + \overset{**}{\varPhi}[f_2] + \overset{**}{\varPhi}[f_1]\overset{**}{\varPhi}[f_2].$$

---

[1] Cf. *C. R. de l'Acad. des Sciences*, Vol. 124, 1897.

In general, we shall have therefore that the new functions of z obtained by means of (3) from known integral functions $\varphi(z)$ (or from known meromorphic functions) will have properties that are analogous to those of the original functions $\varphi(z)$. Thus, for example, addition theorems still exist, but instead of algebraic addition theorems we now have *integral addition theorems* (of the second kind) (e.g. formula (5)).

105. For composition of the second kind, the functions called *Evans's periodic kernels* are worth special study; i.e. functions of the form $f(x-y) + \varphi(x+y)$, where f and φ are periodic functions with period equal to the interval of integration; G. C. EVANS has shown that their reciprocal kernels (4) are of the same form[1].

---

[1] Cf. EVANS, (10). Cf. also ANDREOLI, (1), (4).

BIBLIOGRAPHY IV.

(1) ANDREOLI (G.).—Sui nuclei periodici di Evans e la composizione di seconda specie. (*R. Acc. dei Lincei. Rend.*, Vol. XXV, 1916. Notes I and II.)

(2) ANDREOLI (G.).—Sulla risoluzione di certe equazioni di composizione di seconda specie. (*R. Acc. dei Lincei. Rend.*, Vol. XXV, 1916. Note I.) Sulla soluzione generale di una classe di equazioni di composizione. (*R. Acc. dei Lincei. Rend.*, Vol. XXV, 1916. Note II.)

(3) ANDREOLI (G.).—Sopra certe equazioni di composizione di seconda specie. (*R. Acc. dei Lincei. Rend.*, Vol. XXVI, 1917.)

(4) ANDREOLI (G.).—Sulle equazioni integrali a nuclei di Evans. Naples. 1923.

(5) BOMPIANI (E.).—Sopra le funzioni permutabili. (*R. Acc. dei Lincei, Rend.*, Vol. XIX, ser. 5a, 1910.)

(6) DANIELE (E.).—Sui nuclei che si riproducono per iterazione (Algebra delle funzioni permutabili). (*Cir. Mat. di Palermo. Rend.*, Vol. XXXVII, 1914.)

(7) EVANS (G. C.).—Sopra l'algebra delle funzioni permutabili. (*R. Acc. dei Lincei. Mem.*, ser. 5a, Vol. VIII, 1911.)

(8) EVANS (G. C.).—Applicazione dell'algebra delle funzioni permutabili al calcolo delle funzioni associate. (*R. Acc. dei Lincei. Rend.*, ser. 5a, Vol. XX, 1911.)

(9) EVANS (G. C.).—L'algebra delle funzioni permutabili e non permutabili. (*Cir. Mat. di Palermo. Rend.*, Vol. XXXIV, 1912.)

(10) EVANS (G. C.).—Application of an equation in variable differences to integral equations. (*Am. Math. Soc. Bull.*, Vol. XXII, 1916.)

(11) HADAMARD (J.).—Recherches sur les solutions fondamentales et l'intégration des équations lineaires aux dérivées partielles. (*Ann. Ec. Norm.*, 3ième ser., Vol. XXI, Paris, 1905.)

(12) HADAMARD (J.).—Le principe de Huygens. (*Soc. Math. de France. Bull.*, Vol. (52) 3-4, 1924.)

(13) LAURICELLA (G.).—Sopra l'algebra delle funzioni permutabili di seconda specie. (*Ann. di Mat.*, ser. 3a, Vol. XXI, 1913).

(14) LAURICELLA (G.).—Sopra le funzioni permutabili di seconda specie. (*R. Acc. dei Lincei, Rend.*, ser. 5a, Vol. XXII, 1913.)

(15) LEBESGUE (H.).—Sur un théorème de M. Volterra. (*Soc. Math. de France. Bull.*, 1912.)

(16) LŪSIS (A.).—Permūtāblas funkcijas un Volterra integrālvienādojums (On permutable functions and Volterra's integral equation). (*Annales of the University of Latvia*, Vol. XVII, 1927.)

136 BIBLIOGRAPHY CHAP.

(17) NALLI (P.).—Sopra una relazione fra la teoria della composizione di prima specie e lo studio delle serie divergenti. (*Cir. Mat. di Palermo. Rend.*, Vol. 42, 1917.)

(18) NÖRLUND (N.-E.).—Sur une application des fonctions permutables. (*Lunds. Univ. Arsskrift.*, N. F. Avd. 2, Vol. 16, No. 3.)

(19) PÉRÈS (J.).—Sulle equazioni integrali (*R. Acc. dei Lincei. Rend.* Ser. 5a, Vol. XXII, 1913, 1° Sem.).

(20) PÉRÈS (J.).—Sur les fonctions permutables analytiques (deux Notes). (*R. Acc. dei Lincei. Rend.*, ser. 5a, Vol. XXII, 1913, 2° Sem.; Vol. XXIII, 1914, 1° Sem.)

(21) PÉRÈS (J.).—Sur les fonctions permutables de première espèce de M. Volterra. (*Jour. de Math.*, 7ème ser., Vol. I, 1915.)

(22) PÉRÈS (J.).—These de doctorat. Sur les fonctions permutables de première espèce de M. Vito Volterra. (Gauthier-Villars, Paris, 1915.) (This paper is a reprint of the preceding paper.)

(23) PÉRÈS (J.).—Sur la composition de première espèce: Les fonctions d'ordre quelconque et leur composition. (2 notes.) (*R. Acc. dei Lincei. Rend.*, ser. 5a, Vol. XXVI, 1917.)

(24) PÉRÈS (J.).—Quelques propriétés des fonctions de Bessel (deux Notes). (*R. Acc. dei Lincei. Rend.*, Vol. 27, 1918, 1° Sem.)

(25) PÉRÈS (J.).—Sur certaines transformations fonctionnelles. (*C. R. de l'Acad. des Sciences de Paris*, Vol. 166, 1918.)

(26) PÉRÈS (J.).—Sur certaines transformations fonctionnelles et leur application à la theorie des fonctions permutables. (*Ann. Ec. Norm.*, Vol. XXXVI, Paris, 1919.)

(27) PÉRÈS (J.).—Sur les transformations qui conservent la composition. (*Soc. Math. de France. Bull.*, Vol. XLVII, 1919.)

(28) PÉRÈS (J.).—Sulla teoria delle funzioni permutabili. (*Seminario Mat. Rend.*, Vol. VI, Roma, 1920.)

(29) PÉRÈS (J.).—Transformations qui conservent la composition. Sur les fonctions permutables. (2 notes.) (*R. Acc. dei Lincei. Rend.*, ser. 5a, Vol. XXX, 1921.)

(30) PÉRÈS (J.).—Sur l'inversion de certaines relations intégrales. (*Comp. Rend. Congrès des Soc. Savants*, 1922.)

(31) PÉRÈS (J.).—Quelques compléments sur les transformations qui conservent la composition. (*R. Acc. dei Lincei. Rend.*, Ser. 5a, Vol. XXXIII, 1924, 2° Sem.)

(32) RECCHIA (M.).—Sopra l'algebra delle funzioni permutabili di seconda specie. (*Atti Acc. Giœnia-Catania*, ser. 5a, Vol. X, 1917.)

(33) SEVERINI (C.).—Sulle funzioni permutabili di seconda specie. (*Atti Acc. Giœnia-Catania*, ser. 5a, Vol. VII, 1914.)

(34) SINIGALLIA (L.).—Sulle funzioni permutabili di seconda specie (4 notes). (*R. Acc. dei Lincei. Rend.*, ser. 5a, Vol. XX, 1911; Vol. XXI, 1912; Vol. XXII, 1913.)

(35) VESSIOT (E.).—Sur les groupes fonctionnels et les équations intégrodifferentielles linéaires. (*C. R. de l'Acad. des Sciences de Paris*, Vol. 154, 1912.)

(36) VESSIOT (E.).—Sur les fonctions permutables et les groupes continus de transformations fonctionnelles linéaires. (*C. R. de l'Acad. des Sciences de Paris*, Vol. 154, 1912.)

IV BIBLIOGRAPHY 137

(37) VOLTERRA (V.).—Questioni generali sulle equazioni integrali ed integro-differenziali. (*R. Acc. dei Lincei*, ser. 5a, Vol. XIX, 1910.)
(38) VOLTERRA (V.).—Sopra le funzioni permutabili. (*R. Acc. dei Lincei. Rend.*, ser. 5a, Vol. XIX, 1910.)
(39) VOLTERRA (V.).—Sopra le funzioni permutabili di seconda specie e le equazioni integrali. (*R. Acc. dei Lincei. Rend.*, ser. 5a, Vol. XX, 1911.)
(40) VOLTERRA (V.).—Contributo allo studio delle funzioni permutabili. (*R. Acc. dei Lincei. Rend.*, ser. 5a, Vol. XX, 1911.)
(41) VOLTERRA (V.).—The theory of permutable functions. (Lectures delivered at Princeton University, October 1912.) (Louis Clark Vonuxem Foundation, Princeton Univ. Press, 1915.)
(42) VOLTERRA (V.).—Teoria delle potenze, dei logaritmi e delle funzioni di composizione. (*R. Acc. dei Lincei. Mem.*, ser. 5a, Vol. XI, 1916.)
(43) VOLTERRA (V.).—Functions of composition. (*Rice Inst. Pamph.*, Vol. VII, No. 4, 1920.)
(44) VOLTERRA (V.) and PÉRÈS (J.).—Leçons sur la composition et les fonctions permutables. (Coll. Borel.) Paris, Gauthier-Villars, 1924.
(45) VOLTERRA (V.).—Sur les fonctions permutables. (*Soc. Math. de France. Bull.*, Vol. LII [3-6], 1924.)

CHAPTER V

INTEGRO-DIFFERENTIAL EQUATIONS AND FUNCTIONAL DERIVATIVE EQUATIONS

Section I. Integro-Differential Equations.

§ 1. *Generalities on Integro-Differential Equations.*

106. In general, *differential* equations are equations containing relations of the ordinary type between the unknown functions, their derivatives, and the independent variables. *Integral* equations (cf. Chapter II), on the other hand, are those in which the unknown functions appear under the sign of integration.

We shall now apply the term *integro-differential equations* to those in which the unknown functions appear with their derivatives, and either the unknown functions, or their derivatives, or both, appear under the sign of integration. This, however, is a purely formal classification, since we can easily pass from one type to the other.

Thus, for example, functional equations of the type

$$F[y(t); x] = z(x), \qquad (1)$$

which we can suppose derived from a system of equations in finite terms of the type

$$f_i(y_1, y_2, \ldots y_n) = z_i \quad (i = 1, 2, \ldots n), \qquad (2)$$

by the usual process of passing from a finite to an infinite number of variables, are classified as *integral equations* (cf. No. 41), since we suppose, in general, that the functional $F[y(t); x]$ can be expanded in a series of integrals analogous to TAYLOR's series (No. 30). If instead of systems of equations (2) in finite terms we consider systems of ordinary differential equations of the type

V INTEGRO-DIFFERENTIAL EQUATIONS

$$\begin{aligned}\frac{dy_1}{dx} &= f_1(y_1, y_2, \ldots y_n, x) \\ \frac{dy_2}{dx} &= f_2(y_1, y_2, \ldots y_n, x) \\ &\cdots\cdots\cdots\cdots\cdots \\ \frac{dy_n}{dx} &= f_n(y_1, y_2, \ldots y_n, x)\end{aligned}\Bigg\}, \qquad (3)$$

the usual method, already applied several times, of passing from a finite to an infinite number of variables by substituting the parameter t varying continuously in the interval (a, b) for the discontinuous indices $1, 2, \ldots n$, leads to equations of the form

$$\frac{\partial y(x, \xi)}{\partial x} = \underset{a}{\overset{b}{F}}[y(x, t); x, \xi](^1) \qquad (4)$$

in the unknown function $y(x, \xi)$; and if we suppose that the functional F can also be expanded in a series of integrals analogous to TAYLOR's series (cf. Chapter I, No. 30), equations of this kind will have to be considered, according to our definitions, as *integro-differential equations*; we shall call them *ordinary* integro-differential equations *of the first order*.

107. It is however easy to see that these equations of the type (4) can be reduced immediately to integral equations of the type (1) studied at an earlier stage. We have in fact from (4), integrating from x_0 to x, and denoting by $y_0(\xi)$ the (arbitrary) value of the unknown $y(x, \xi)$ for $x = x_0$ (the initial value),

$$y(x, \xi) = y_0(\xi) + \int_{x_0}^x \underset{a}{\overset{b}{F}}[y(x, t); x, \xi]\, dx, \qquad (5)$$

an equation which, if the first variable x of the unknown function is considered as a constant parameter, becomes identical with the equations of the type (1), whose conditions of solubility have already been studied in Chapter II (Nos. 54 to 56). Further, for the integration of equation (4) we can proceed directly by PICARD's method of successive approximations. If in fact we suppose that the functional F is continuous, and that we have always

$$|\,F[y_1(t); x, \xi] - F[y_2(t); x, \xi]\,| < A\,\varepsilon,$$

(¹) For the notation cf. No. 6.

140 THEORY OF FUNCTIONALS CHAP.

A being a constant, provided that

$$|y_1(t) - y_2(t)| < \varepsilon$$

(this is analogous to LIPSCHITZ's condition for ordinary differential systems), then, given arbitrarily the initial value $\psi(\xi) = y_0(\xi) = y(x_0, \xi)$ of the unknown function for $x = x_0$, we can construct a sequence of functions $y_i(x, \xi)$ which can be shown to converge uniformly towards a function $y(x, \xi)$ depending on $\psi(\xi)$ (a functional of ψ) given by

$$y(x, \xi) = \psi(\xi) + \Phi[\psi(t); x, \xi],$$

which constitutes, with the foregoing hypotheses, the unique solution of (4) that takes for $x = x_0$ the value $\psi(\xi)$ fixed in advance.

108. It is easy to see the difficulty of a rational classification of integro-differential equations. *If the derivatives are always taken with respect to a single variable*, the integro-differential equations are called *ordinary*, and of order n if n is the order of the highest derivative. Other integro-differential equations, on the contrary, which often occur in questions of physics and mathematics (Nos. 116 to 121), contain *derivatives with respect to different variables*; these equations are therefore called *partial derivative integro-differential equations*. We shall give some examples below in §§ 3 to 7. Partial derivative integro-differential equations have characteristic differences from ordinary integro-differential equations, as in the case of common differential equations, but it may be pointed out that there are substantial differences even between the various types of ordinary integro-differential equations. Thus, for example, while the solution of the ordinary equations considered in Nos. 106, 107 and 109 depends on arbitrary *functions*, the solution of the equations, also ordinary, considered in No. 115 depends only on arbitrary *constants*.

There is a whole large class of integro-differential equations which are closely connected with differential equations and can be classified like them. They are precisely all those obtained from the theory of permutable functions by a method analogous to that used for integral equations in Nos. 84 to 86; we shall proceed to consider these equations in § 3 and in No. 118.

We can, among partial derivative integro-differential equations, pick out some types which are analogous to the elliptic, hyperbolic and parabolic types of differential equations. We shall discuss these integro-differential equations in §§ 6 and 7.

The essential question of the existence of the integrals of integro-differential equations has been discussed by L. POMEY, in both the linear and non-linear cases, for any number of variables and of equations, even when this number is infinite. Further, L. POMEY's results provide solutions of integrodifferential equations, sometimes for all values of the variables, and sometimes in finite domains.([1])

§ 2. *First Examples of Integro-Differential Equations.*

109. A particularly interesting case is one studied by SCHLESINGER ([2]), in which the functional F is a linear functional, i.e. the equation (4) is of the type

$$\frac{\partial y(x, \xi)}{\partial x} = \int_a^b f(x, \xi, t) y(x, t) dt.$$

For these equations a theory can be constructed which is completely analogous to that of FUCHS for systems of linear differential equations.

Other more general linear integro-differential equations have been discussed by L. POMEY (cf. No. 108).

110. Other interesting examples of integro-differential equations are obtained from the *calculus of variations* by equating to zero the functional derivative of a functional F in order to determine the maxima or minima of F (cf. No. 33). Thus, for example, for the functional

$$F[y(t)] = \int_a^b \int_a^b f(t, u, y(t), y(u)) dt du,$$

on equating to 0 the functional derivative we get the *integral* equation

$$F'[y(t); x] = \int_a^b (f_1 + f_2) dt = 0,$$

([1]) Cf. L. POMEY, (45) to (51).
([2]) Cf. SCHLESINGER, (53), (54).

142 THEORY OF FUNCTIONALS CHAP.

where
$$\frac{\partial f(x, t, y(x), y(t))}{\partial y(x)} = f_1, \quad \frac{\partial f(t, x, y(t), y(x))}{\partial y(x)} = f_2.$$

For the functional
$$F[y(t)] = \int_a^b \int_a^b f(t, x, y(t), y(x), y'(t), y'(x)) \, dt \, dx,$$
however, putting
$$f_1 = \frac{\partial f}{\partial y(t)}, \quad f_2 = \frac{\partial f}{\partial y(x)}, \quad f_3 = \frac{\partial f}{\partial y'(t)}, \quad f_4 = \frac{\partial f}{\partial y'(x)},$$
we shall obtain
$$\delta F = \int_a^b \delta y(x) \, dx \int_a^b (\bar{f}_1 + f_2) \, dt + \int_a^b \delta y'(x) \, dx \int_a^b (\bar{f}_3 + f_4) \, dt,$$

where $\bar{f}_1, \bar{f}_3$ denote the expressions obtained from f_1 and f_3 respectively by interchanging t and x. Integrating the second term by parts, we get for the derivative $F'[y(t); x]$ the expression
$$F'[y(t); x] = \int_a^b (\bar{f}_1 + f_2) \, dt - \frac{d}{dx} \int_a^b (\bar{f}_3 + f_4) \, dt.$$

If we equate this expression to 0, or to a known function $z(x)$, we shall evidently get an *integro-differential* equation (since f_1, $f_2, \bar{f}_3, f_4$ contain y and its derivative), of a different type, however, from that considered in No. 106, in which the variable x of differentiation (of the unknown function $y(x, t)$) was distinct from the variable t of integration.

111. Another example of an *integro-differential* equation which is also reducible to an *integral* equation has already occurred and been solved in No. 89 above; this is the equation
$$\frac{\partial \varphi}{\partial x} + \frac{\partial \varphi}{\partial y} = \int_x^y [\varphi(x, \xi) F(\xi, y) - F(x, \xi) \varphi(\xi, y)] \, d\xi$$

in the unknown function $\varphi(x, y)$, which we met in trying to determine all the functions which are permutable of the first kind with a given function $f(x, y)$.

§ 3. Integro-Differential Equations obtained from the Theory of Permutable Functions of the First Kind.

112. Let us suppose that we have an algebraic relation

$$\Phi\left(z_1, z_2, \ldots z_n; \varphi, \frac{\partial \varphi}{\partial z_1}, \frac{\partial \varphi}{\partial z_2}, \ldots \frac{\partial^{p_1+p_2+\cdots+p_n}\varphi}{\partial z_1^{p_1} \partial z_2^{p_2} \ldots \partial z_n^{p_n}}, \ldots\right) = 0 \quad (1)$$

connecting $z_1, z_2, \ldots z_n$, the function $\varphi(z_1, z_2, \ldots z_n)$, and its derivatives up to a certain order; and that

$$\varphi(z_1, z_2, \ldots z_n) = \sum a_{i_1 i_2 \ldots i_n} z_1^{i_1} z_2^{i_2} \ldots z_n^{i_n} \quad (2)$$

is a solution of the *differential equation* (1). If now we substitute $z_1 \xi_1, z_2 \xi_2, \ldots z_n \xi_n$ for $z_1, z_2, \ldots z_n$ respectively, and $\dfrac{\psi}{\xi_0}$ for φ, where the ξ's are independent of the z's, the equation (1) will become

$$\Phi\left(z_1\xi_1, \ldots z_n\xi_n; \frac{\psi}{\xi_0}, \frac{1}{\xi_0\xi_1}\frac{\partial \psi}{\partial z_1}, \ldots \frac{1}{\xi_0\xi_1^{p_1}\ldots\xi_n^{p_n}}\frac{\partial^{p_1+\cdots+p_n}\psi}{\partial z_1^{p_1}\ldots\partial z_n^{p_n}}\right) = 0,$$

or, reducing to the integral form,

$$\Psi\left(z_1, z_2, \ldots z_n; \xi_0, \xi_1, \ldots \xi_n; \psi, \frac{\partial \psi}{\partial z_1}, \frac{\partial \psi}{\partial z_2}, \ldots\right) = 0, \quad (3)$$

a differential equation which will be satisfied identically by

$$\psi = \xi_0 \varphi(z_1\xi_1, z_2\xi_2, \ldots z_n\xi_n). \quad (4)$$

If now for $\xi_0, \xi_1, \ldots \xi_n$ we substitute a set of functions $f_0(x, y)$, $f_1(x, y), \ldots f_n(x, y)$ which are arbitrary but permutable with each other (of the first kind), and if in (3) and (4) we interpret the ordinary products as *products by composition* of the first kind, then the functions Ψ and ψ become *functions by composition* of the f's, and from equation (3) we get the *integro-differential* equation

$$\overset{*}{\Psi}\left(z_1, z_2, \ldots z_n; \overset{*}{f_0}, \overset{*}{f_1}, \ldots \overset{*}{f_n}; \psi, \frac{\partial \overset{*}{\psi}}{\partial z_1}, \frac{\partial \overset{*}{\psi}}{\partial z_2}, \ldots\right) = 0 \quad (5)$$

in the unknown function $\psi(z_1, z_2, \ldots z_n; x, y)$; an equation for which we know at once one solution given by

$$\psi(z_1, z_2, \ldots z_n; x, y) = \overset{*}{f_0} \varphi(z_1 \overset{*}{f_1}, z_2 \overset{*}{f_2}, \ldots z_n \overset{*}{f_n}). \quad (6)$$

We have therefore the following general theorem on integro-differential equations:

If the integro-differential equation (5) *can be obtained in the manner just described from a differential equation* (1) *for which an analytical and regular solution φ is known, then the equation* (5) *will have for solution a function ψ, given by* (6), *which is an integral function of* $z_1, z_2, \ldots z_n$ ([1]) (cf. No. 84).

These considerations can be immediately extended to the case of several differential equations with several unknown functions. This case gives rise to *systems of integro-differential equations*, for which we shall know a system of solutions whenever we know a system of solutions for the original differential equations.

We have also to remember the results of No. 83 about integral addition theorems, which can be applied to the functions obtained in the way indicated above when there are addition theorems for the integrals of the corresponding differential equations.

113. As an application of the above results let us consider the ordinary differential system satisfied by the three elliptic functions sn, cn, dn,

$$\frac{d}{dz}\operatorname{sn} z = \operatorname{cn} z \operatorname{dn} z,$$

$$\frac{d}{dz}\operatorname{cn} z = -\operatorname{sn} z \operatorname{dn} z,$$

$$\frac{d}{dz}\operatorname{dn} z = -k^2 \operatorname{sn} z \operatorname{cn} z.$$

Putting $\varphi_1 = \xi \operatorname{sn} \xi z, \quad \varphi_2 = \xi \operatorname{cn} \xi z, \quad \varphi_3 = \xi \operatorname{dn} \xi z,$
we shall have

$$\frac{d\varphi_1}{dz} = \varphi_2 \varphi_3, \quad \frac{d\varphi_2}{dz} = -\varphi_3 \varphi_1, \quad \frac{d\varphi_3}{dz} = -k^2 \varphi_1 \varphi_2, \quad (7)$$

and $\varphi_1, \varphi_2, \varphi_3$ can be expanded, in the neighbourhood of $\xi = 0$, $z = 0$, in power series of the type

$$\left.\begin{array}{l}\varphi_1 = a_1 \xi^2 z + a_2 \xi^4 z^3 + \cdots \\ \varphi_2 = b_1 \xi \phantom{{}+{}} + b_2 \xi^3 z^2 + \cdots \\ \varphi_3 = c_1 \xi \phantom{{}+{}} + c_2 \xi^3 z^2 + \cdots\end{array}\right\}. \quad (8)$$

([1]) Cf. VOLTERRA, (61).

V   INTEGRO-DIFFERENTIAL EQUATIONS   145

If now we substitute $f(x, y)$ (an arbitrary function) for ξ, and instead of the ordinary powers consider the *powers by composition*, the three series (2) become

$$\left.\begin{aligned}\varphi_1(z;x,y) &= a_1 \overset{*}{f^2} z + a_2 \overset{*}{f^4} z^3 + \cdots \\ \varphi_2(z;x,y) &= b_1 f \;\; + b_2 \overset{*}{f^3} z^2 + \cdots \\ \varphi_3(z;x,y) &= c_1 f \;\; + c_2 \overset{*}{f^3} z^2 + \cdots\end{aligned}\right\}, \quad (9)$$

which represent three *integral functions of z which satisfy the system of integro-differential equations*

$$\left.\begin{aligned}\frac{d\varphi_1}{dz} &= \int_x^y \varphi_2(z;x,\xi)\varphi_3(z;\xi,y)\,d\xi \\ \frac{d\varphi_2}{dz} &= -\int_x^y \varphi_3(z;x,\xi)\varphi_1(z;\xi,y)\,d\xi \\ \frac{d\varphi_3}{dz} &= -k^2 \int_x^y \varphi_1(z;x,\xi)\varphi_2(z;\xi,y)\,d\xi\end{aligned}\right\}, \quad (10)$$

a set of non-linear equations belonging to a new type of integro-differential equations. The solutions (9) of this system will thus have *integral addition theorems* (cf. No. 83) with respect to the variable z, which will be obtained from the corresponding algebraic addition theorems for the elliptic functions.

§ 4. *Integro-Differential Equations obtained from the Theory of Permutable Functions of the Second Kind.*

114. In addition to the class just considered of integro-differential equations, obtained from differential equations by substituting *compositions of the first kind* of the f's for the ordinary products of the z's, we can consider a further class obtained in a similar way by substituting *compositions of the second kind*. We have thus that, given the differential equation

$$\Phi\left(z_1, z_2, \ldots z_n; \varphi, \frac{\partial \varphi}{\partial z_1}, \frac{\partial \varphi}{\partial z_2}, \ldots \frac{\partial^{p_1 + p_2 + \cdots + p_n}\varphi}{\partial z_1^{p_1} \partial z_2^{p_2} \ldots \partial z_n^{p_n}}, \ldots\right) = 0, \quad (1)$$

a solution $\varphi(z_1, z_2, \ldots z_n)$ of which is known, we can obtain from it the equation

$$\Psi\left(z_1, z_2, \ldots z_n; \xi_0, \xi_1, \ldots \xi_n; \psi, \frac{\partial \psi}{\partial z_1}, \frac{\partial \psi}{\partial z_2}, \ldots\right) = 0 \quad (2)$$

(cf. No. 112) with the solution $\psi = \xi_0 \varphi(z_1 \xi_1, z_2 \xi_2, \ldots z_n \xi_n)$; whence it follows that *the integro-differential equation*

$$\overset{**}{\Psi}\left(z_1, z_2, \ldots z_n; \overset{**}{f_0}, \overset{**}{f_1}, \ldots \overset{**}{f_n}; \psi, \frac{\partial \overset{**}{\psi}}{\partial z_1}, \frac{\partial \overset{**}{\psi}}{\partial z_2}, \ldots\right) = 0, \quad (3)$$

obtained from (2) by substituting $f_i(x, y)$ for ξ_i, and compositions of the f's of the second kind for the products of the ξ's, will have for its solution

$$\psi(z_1, z_2, \ldots z_n; x, y) = \overset{**}{f_0} \overset{**}{\varphi} (z_1 \overset{**}{f_1}, z_2 \overset{**}{f_2}, \ldots z_n \overset{**}{f_n}).$$

This function ψ, a solution of the integro-differential equation (3), will be an integral function in $z_1, z_2, \ldots z_n$ if $\varphi (v_1, v_2, \ldots v_n)$, the solution of the differential equation (1), was an integral function, and will be the quotient of two integral functions of $z_1, z_2, \ldots z_n$ *(meromorphic)* if φ was such a quotient (cf. Nos. 103 to 105).

We can repeat here what was said in No. 112 (cf. No. 83) about integral addition theorems.

The same considerations will hold for systems of differential equations, which will be transformed into *systems of integro-differential equations* by substituting compositions of the second kind for ordinary products. Thus, for example, from the series (8) of No. 113, which are meromorphic functions of z, we get the functions

$$\overline{\varphi}_1(z; x, y) = a_1 \overset{**}{f^2} z + a_2 \overset{**}{f^4} z^3 + \cdots,$$

$$\overline{\varphi}_2(z; x, y) = b_1 f + b_2 \overset{**}{f^3} z^2 + \cdots,$$

$$\overline{\varphi}_3(z; x, y) = c_1 f + c_2 \overset{**}{f^3} z^2 + \cdots,$$

($f(x, y)$ an arbitrary function), which are also meromorphic functions of z, and will satisfy the system of integro-differential equations

$$\frac{\partial \overline{\varphi}_1(z; x, y)}{\partial z} = \int_a^b \overline{\varphi}_2(z; x, \xi) \overline{\varphi}_3(z; \xi, y) d\xi,$$

$$\frac{\partial \overline{\varphi}_2(z; x, y)}{\partial z} = -\int_a^b \overline{\varphi}_3(z; x, \xi) \overline{\varphi}_1(z; \xi, y) d\xi,$$

$$\frac{\partial \overline{\varphi}_3(z; x, y)}{\partial z} = -k^2 \int_a^b \overline{\varphi}_1(z; x, \xi) \overline{\varphi}_2(z; \xi, y) d\xi,$$

V INTEGRO-DIFFERENTIAL EQUATIONS 147

which differ from the system (10) of No. 113 only in having the limits of the integrals constant.

We thus see how from the differential equations of the elliptic functions we can obtain two different types of integro-differential systems: the first, with the limits of the integrals variable, which has integral functions of z for solutions; the second, that just examined, with the limits of the integrals constant, which has meromorphic functions of z for solutions. It is also to be noted that all the integral functions of z obtained in this way are represented by *very rapidly convergent series*, which are therefore very well adapted for actual calculation. In the two cases the addition theorems for the elliptic functions lead to integral addition theorems.

§ 5. *Preliminary Considerations on Hereditary Questions. Integro-Differential Equations for the Elastic Torsion of a Wire.*

115. When we take account of hereditary phenomena, the questions of physics and mechanics lead to integro-differential equations. We shall discuss this subject in the next chapter, and here we shall give only a first sketch strictly connected with integro-differential equations.

A hereditary phenomenon occurs in a system when the phenomenon does not depend only on the actual state of the system or on its immediately preceding states (i.e. on the initial values of the parameters which define the state of the system and on some of their derivatives with respect to the time) but on all the preceding states through which the system has passed; that is to say, it is a phenomenon which depends on the previous *history* of the system, and may therefore be called *hereditary*. Examples are elastic phenomena, in which the deformation of an elastic bar or the torsion of a wire does not depend only on the nature of the forces applied, but also on the previous deformations undergone by the bar or the wire; others, again, are the phenomena of *magnetic hysteresis, electric hysteresis, lag*, etc.

As we wish to study in particular the case of the *torsion of a wire*, we may recall the result that to a first approximation the connection between the moment m of the torsional couple and the corresponding angle of torsion ω is given, in the case of statical equilibrium, by the linear relations

$$\omega = km, \quad m = h\omega, \quad h = \frac{1}{k}, \tag{1}$$

where k and $h = \frac{1}{k}$ are constants depending on the characteristics of the wire. Accurate experiments show, however, that ω does not depend only on the actual torsional moment but also on all the preceding moments; i.e. the phenomenon is *hereditary*. If $m(\tau)$ denotes the torsional moment acting on the wire at time τ, it follows that in order to find the angle of torsion $\omega(t)$ at time t we must add to the right-hand side of the first equation (1) a corrective term, depending on all the values of $m(\tau)$ for τ prior to t, and therefore a *functional* $F\left[\underset{-\infty}{\overset{t}{m(\tau)}}\right]$ of $m(\tau)$; thus the equation in question takes the more precise form

$$\omega = km(t) + F\left[\underset{-\infty}{\overset{t}{m(\tau)}}\right]. \tag{2}$$

If the functional F can be expanded in a series analogous to TAYLOR's series (No. 30) and if all the terms in it of higher degree than the first can be neglected (linear heredity), (2) reduces to a linear relation

$$\omega(t) = km(t) + \int_{-\infty}^{t} f(t,\tau) m(\tau) d\tau \tag{3}$$

between $\omega(t)$ and $m(t)$. Solving this VOLTERRA's integral equation (No. 43) with respect to $m(t)$ and denoting the reciprocal kernel of $\frac{1}{k} f(t,\tau)$ by $\varphi(t,\tau)$, we get

$$m(t) = h\omega(t) + \int_{-\infty}^{t} \varphi(t,\tau) \omega(\tau) d\tau. \tag{4}$$

If we suppose that the hereditary effects prior to an instant $t_0 < t$ are negligible, where for simplicity we can consider t_0 as the origin of the time ($t_0 = 0$), then (4) becomes

$$m(t) = h\omega(t) + \int_{0}^{t} \varphi(t,\tau) \omega(\tau) d\tau, \tag{5}$$

in which both the limits of the integral are finite.

If from the statical case discussed above we pass to the *dynamical* case, and try to study the *oscillations* of the wire, i.e. if we suppose in general that the angular velocity and acceleration are no longer negligible, the equation of motion of the wire will be obtained from (5) by means of d'Alembert's principle, by substituting $m(t) - \mu \dfrac{d^2\omega}{dt^2}$ (μ constant) for $m(t)$; we shall thus get

$$m(t) - \mu \frac{d^2\omega}{dt^2} = h\omega(t) + \int_0^t \varphi(t,\tau)\omega(\tau)\,d\tau. \qquad (6)$$

If then we know the moment $m(t)$ at each instant we have, to determine the angle $\omega(t)$, the *integro-differential equation* (6) *of the second order*, in which the unknown $\omega(t)$ occurs under the integral sign and with its derivatives. To solve it we have only to integrate twice with respect to t from 0 to t, which gives the *Volterra's integral equation of the second kind* (No. 43)

$$\mu\omega(t) + \int_0^t \psi(t,\tau)\omega(\tau)\,d\tau = \chi(t) + \mu\omega(0) + \mu\omega'(0)t, \qquad (7)$$

in which

$$\psi(t,\tau) = \int_\tau^t \left[h + \int_\tau^\vartheta \varphi(\xi,\tau)\,d\xi\right]d\vartheta,$$

$$\chi(t) = \int_0^t d\vartheta \int_0^\vartheta m(\xi)\,d\xi,$$

and from the integral equation (7), when the two arbitrary constants $\omega(0)$ and $\omega'(0)$ are known, we can get the solution $\omega(t)$ in the form

$$\omega(t) = \frac{1}{\mu}\left[\chi(t) + \mu\omega(0) + \mu\omega'(0)t\right]$$
$$+ \int_0^t \overline{\psi}(t,\tau)[\chi(\tau) + \mu\omega'(0)\tau + \mu\omega(0)]\,d\tau,$$

where $\overline{\psi}$ is the solvent of ψ ([1]).

§ 6. *Integro-Differential Equations of Elliptic Type.*

116. Another integro-differential equation, which we now consider as a *partial derivative* equation *of elliptic type*, and which occurs in questions of hereditary phenomena in physics,

---
[1] Cf. GRAFFI, (74).

is the following:

$$\Delta^2 u(t) + \int_0^t \left[\frac{\partial^2 u(\tau)}{\partial x^2} f(t,\tau) \right.$$
$$\left. + \frac{\partial^2 u(\tau)}{\partial y^2} \varphi(t,\tau) + \frac{\partial^2 u(\tau)}{\partial z^2} \psi(t,\tau) \right] d\tau = 0 \qquad (1)$$

where we have put

$$\Delta^2 u(t) = \frac{\partial^2 u(x,y,z,t)}{\partial x^2} + \frac{\partial^2 u(x,y,z,t)}{\partial y^2} + \frac{\partial^2 u(x,y,z,t)}{\partial z^2}.$$

Considering x, y, z as the cartesian co-ordinates of a point in a region S of space bounded by a surface σ, it can be shown [1] that *every solution u of the equation (1) which is regular (i.e. finite and continuous, together with its first and second derivatives with respect to x, y, z) is determinate in the interior of S for values of t between 0 and T if for the same values of t the values of u on the boundary σ of S are known.*

117. In order actually to construct a solution u by means of the elements that determine it, we can proceed by means of a method analogous to that used for LAPLACE's equation. For this equation, we prove first a reciprocity theorem between two of its solutions u and v, expressed by

$$\int_\sigma u \frac{dv}{dn} d\sigma = \int_\sigma v \frac{du}{dn} d\sigma$$

(where n denotes the normal to σ), and by means of the particular solution $\frac{1}{r}$ (the fundamental solution), where r is the distance between the point x, y, z and a point A within S, we get the value of the solution at the pole A at which $\frac{1}{r}$ becomes infinite. Similarly, in the case of equation (1), we can first prove a *reciprocity theorem* and then construct a *fundamental solution*. More precisely, if $v(x, y, z, t)$ is a solution of the integro-differential equation

$$\Delta^2 v(t) + \int_t^\vartheta \left[\frac{\partial^2 v(\tau)}{\partial x^2} f(\tau,t) \right.$$
$$\left. + \frac{\partial^2 v(\tau)}{\partial y^2} \varphi(\tau,t) + \frac{\partial^2 v(\tau)}{\partial z^2} \psi(\tau,t) \right] d\tau = 0, \qquad (2)$$

---

[1] Cf. VOLTERRA, (64).

which we shall call *adjoint* to (1), the reciprocity theorem is obtained by saying that the expression

$$H_\sigma[u,v;\vartheta] = \int_0^\vartheta dt \int_\sigma \left[v(t) \frac{\partial u(t)}{\partial n} - u(t) \frac{\partial v(t)}{\partial n} \right] d\sigma$$
$$+ \int_0^\vartheta dt \int_t^\vartheta d\tau \int_\sigma \left\{ \left[v(\tau) \frac{\partial u(t)}{\partial x} - u(t) \frac{\partial v(\tau)}{\partial x} \right] f(\tau,t) \cos nx \right.$$
$$+ \left[v(\tau) \frac{\partial u(t)}{\partial y} - u(t) \frac{\partial v(\tau)}{\partial y} \right] \varphi(\tau,t) \cos ny$$
$$+ \left. \left[v(\tau) \frac{\partial u(t)}{\partial z} - u(t) \frac{\partial v(\tau)}{\partial z} \right] \psi(\tau,t) \cos nz \right\} d\sigma \qquad (3)$$

(a functional of the two functions u and v) is identically zero whenever u is a solution of (1) and v of (2) ([1]).

118. If we suppose next that the three functions f, φ, ψ are permutable of the first kind, the equation (2) can be considered as being obtained in the way indicated in No. 112 from the differential equation

$$\frac{\partial^2 v}{\partial x^2}(1+z_1) + \frac{\partial^2 v}{\partial y^2}(1+z_2) + \frac{\partial^2 v}{\partial z^2}(1+z_3) = 0, \qquad (4)$$

by substituting f, φ, ψ for z_1, z_2, z_3 respectively, and compositions of the first kind for their products. Putting

$$1 + z_i = \frac{1}{1-\zeta_i}, \quad \zeta_i = 1 - \frac{1}{1+z_i}, \quad (i = 1, 2, 3) \qquad (5)$$

the fundamental solution of (4), with the pole at the origin, is given by

$$\bar{v} = \frac{\zeta}{\sqrt{x^2(1-\zeta_1) + y^2(1-\zeta_2) + z^2(1-\zeta_3)}}$$

(where ζ is any quantity whatever, independent of x, y, z); or, expanding in series, with $r = \sqrt{x^2 + y^2 + z^2}$,

$$\bar{v}r = \zeta\{1 + k\}, \text{ where } k \text{ is}$$

$$\sum_1^\infty \frac{\frac{1}{2}\left(\frac{1}{2}+1\right)\cdots\left(\frac{1}{2}+n-1\right)}{n!} \left[\left(\frac{x}{r}\right)^2 \zeta_1 + \left(\frac{y}{r}\right)^2 \zeta_2 + \left(\frac{z}{r}\right)^2 \zeta_3 \right]^n, \qquad (6)$$

([1]) Cf. VOLTERRA, (64).

in which, by (5),

$$\zeta_i = z_i - z_i^2 + z_i^3 - \cdots \qquad (i = 1, 2, 3). \qquad (7)$$

From the series (6) and (7) we shall then get, by the usual method of No. 112, a solution of (2), which, as it is obtained from the fundamental solution of (4), will be called the *fundamental solution* of (2), and like the other will become infinite at the origin like $\dfrac{1}{r}$. For this we have only to substitute f, φ, ψ for z_1, z_2, z_3, a function χ permutable with the first three for ζ, and powers by composition for ordinary powers; the series (7) thus become

$$\bar{f}(\tau, t) = f(\tau, t) - \overset{*}{f^2}(\tau, t) + \overset{*}{f^3}(\tau, t) - \cdots,$$
$$\bar{\varphi}(\tau, t) = \varphi(\tau, t) - \overset{*}{\varphi^2}(\tau, t) + \overset{*}{\varphi^3}(\tau, t) - \cdots,$$
$$\bar{\psi}(\tau, t) = \psi(\tau, t) - \overset{*}{\psi^2}(\tau, t) + \overset{*}{\psi^3}(\tau, t) - \cdots;$$

and putting

$$\varrho(x, y, z; \tau, t) = \left(\frac{x}{r}\right)^2 \bar{f}(\tau, t) + \left(\frac{y}{r}\right)^2 \bar{\varphi}(\tau, t) + \left(\frac{z}{r}\right)^2 \bar{\psi}(\tau, t),$$

the expression (6) will give finally the fundamental solution v of (2) in the form

$$v = \frac{\chi(\vartheta, t) + \lambda}{r}$$

where

$$\lambda = \sum_{1}^{\infty} \frac{\frac{1}{2}\left(\frac{1}{2} + 1\right) \cdots \left(\frac{1}{2} + n - 1\right)}{n!} \int_{t}^{\vartheta} \chi(\vartheta, \tau) \overset{*}{\varrho^n}(\tau, t) d\tau, \qquad (8)$$

depending on the arbitrary function $\chi(\vartheta, t)$, which can however be put equal to unity ([1]).

119. Once in possession of the reciprocity theorem (3) and of the fundamental solution (8), there is no difficulty in continuing the application of GREEN's method. If, in fact, in (3) v is precisely the *fundamental* solution (8) of (2), and if the pole at which v becomes infinite is outside the space S, then the reciprocity theorem will certainly be satisfied, i.e. $H_\sigma[u, v; \vartheta] = 0$. If

---

([1]) Cf. VOLTERRA, (87) of Bibliography I, p. 153.

however the pole is inside S, the same theorem will no longer be applicable, since v is no longer *regular* in the whole of S (becomes ∞ at the pole). It will however be applicable to that part of S which is obtained from S itself by removing a small sphere with its centre at the pole, bounded by a surface ω, so that the part in question is bounded by σ and ω; and we shall therefore have

$$H_\sigma[u,v;\vartheta] + H_\omega[u,v;\vartheta] = 0.$$

Making the radius of the sphere tend to zero, the foregoing equation will still be satisfied, and since

$$\lim H_\omega[u,v;\vartheta] = -4\pi \int_0^\vartheta u_0(t)\,dt,$$

where $u_0(t)$ is the value of the solution $u(x,y,z,t)$ of (1) at the pole [2], we shall have at the limit

$$4\pi \int_0^\vartheta u_0(t)\,dt = H_\sigma[u,v;\vartheta],$$

$$u_0(\vartheta) = \frac{1}{4\pi} \frac{\partial}{\partial \vartheta} H_\sigma[u,v;\vartheta], \qquad (9)$$

a formula giving the value of the solution u of (1) at a point inside S (the pole of the fundamental solution v) in terms of the values of u and of its derivatives at the boundary σ (values which appear in the expression (3) for H).

If w were another solution of the adjoint equation (2) such that $v+w$ were zero on the boundary σ, we could substitute for (9) the formula

$$u_0(\vartheta) = \frac{1}{4\pi} \frac{\partial}{\partial \vartheta} H_\sigma[u, v+w; \vartheta], \qquad (10)$$

in which, as follows from the expression (3) for H, the value of u at the pole would be given in terms *only of the values of u* on the boundary σ (since with this hypothesis the coefficients of the derivatives of u are zero). This solution $v+w$ of the adjoint equation (2), which vanishes on the boundary σ of a region S, corresponds completely to *Green's function* in the case of LAPLACE's equation.

---

[2] Cf. VOLTERRA, (64); cf. also (87) of Bibliography I, p. 85.

§ 7. *Integro-Differential Equations of Hyperbolic and Parabolic Type.*

120. We have already, in No. 115, considered the equation of the oscillations of an elastic wire. If instead we investigate the vibrations of an elastic cord in the case of linear heredity, we also obtain the partial derivative *integro-differential equation* of hyperbolic type [1]

$$\frac{\partial^2 u(z,t)}{\partial t^2} = \frac{\partial^2 u(z,t)}{\partial z^2} + \int_0^t \frac{\partial^2 u(z,\tau)}{\partial z^2} \psi(t,\tau) d\tau.$$

To integrate it we can, as in the classical case where the term with the integral is absent, apply FOURIER's method of the *separation of the variables;* i.e. we try to find solutions of the form

$$u(z,t) = \sin m(z + \alpha) f(t) \qquad (1)$$

(*m* and α constants). The necessary and sufficient condition for this is that the function $f(t)$ should satisfy the ordinary integro-differential equation

$$\frac{d^2 f(t)}{dt^2} + m^2 \left\{ f(t) + \int_0^t \psi(t,\tau) f(\tau) d\tau \right\} = 0,$$

which we have already solved in No. 115 in the case of the torsional vibrations.

The formula (1) corresponds to Taylor's formula, but we can also obtain a solution of d'Alembert's type for the vibrating string.

121. In other questions, also relating to hereditary phenomena, we arrive at partial derivative integro-differential equations of *parabolic* type, which have been studied by EVANS [2]. These are of the type

$$\frac{\partial u(x,t)}{\partial t} - \frac{\partial^2 u(x,t)}{\partial x^2} - \int_{t_0}^t A(t,\tau) \frac{\partial^2 u(x,\tau)}{\partial x^2} d\tau = 0; \qquad (2)$$

considering this equation as an integral equation of the second kind, and solving it with respect to $\frac{\partial^2 u(x,t)}{\partial x^2}$, it is equivalent to

$$\frac{\partial^2 u(x,t)}{\partial x^2} - \frac{\partial u(x,t)}{\partial t} + \int_{t_0}^t B(t,\tau) \frac{\partial u(x,\tau)}{\partial \tau} d\tau = 0. \qquad (3)$$

---

[1] Cf. VOLTERRA, (65).
[2] Cf. G. C. EVANS, (14).

For these equations EVANS proves that the solution exists and is unique when the values at the limits are given, and gives a general method of solution by successive approximations. If instead we apply the method of the separation of the variables to equation (2), i.e. if we try to find the particular solutions of (2) that are of the type

$$u = f(t)(a \sin kx + b \cos kx),$$

we get for the unknown function $f(t)$ the ordinary integro-differential equation

$$\frac{df}{dt} = -k^2 f(t) - \int_{t_0}^{t} k^2 A(t,\tau) f(\tau) d\tau,$$

which is at once reducible to the integral equation of the second kind

$$f(t) = h - \int_{t_0}^{t} k^2 \left\{ 1 + \int_{\tau}^{t} A(t',\tau) dt' \right\} f(\tau) d\tau,$$

which we know how to solve (No. 43).

Section II. Functional Derivative Equations.

§ 1.

122. We shall now proceed to study a branch of the theory of functionals which is completely new, but is connected with the theory of integro-differential equations, namely, relations between the functional derivatives (cf. Nos. 26 and 27) of a functional, the functional itself, and the independent variables, which we can call *functional derivative equations*. These equations can also be obtained, by the usual process of passing from the finite to the infinite (Nos. 1 to 5), from ordinary *total differential* equations (whose solution depends on arbitrary *constants*) and from *partial differential* equations (whose solution depends on arbitrary *functions*); corresponding to these we shall have two different types of functional derivative equations.

The first example, in order of time, of functional derivative equations was obtained [1] by the extension of the JACOBI-

---

[1] Cf. VOLTERRA, (59).

HAMILTON theory to a particular case of double integrals. In fact, if we try to express the condition that the variation of the integral

$$\iint \left\{ \sum p_{ih} \frac{d(x_i, x_h)}{d(u,v)} - H \right\} du\, dv$$

shall vanish (where as usual $\frac{d(x_i, x_h)}{d(u,v)}$ denotes the functional determinant of x_i and x_h with respect to u and v), we arrive at the partial differential equations

$$\frac{d(x_i, x_s)}{d(u,v)} = \frac{\partial H}{\partial p_{is}}, \qquad \sum_h \frac{d(p_{ih}, x_h)}{d(u,v)} = -\frac{\partial H}{\partial x_i}, \qquad (1)$$

in the three unknown functions x_i of u and v ($i = 1, 2, 3$); equations which offer a certain analogy with the canonical equations of dynamics (cf. also No. 112).

Between these equations (1) and the functional derivative equation

$$H\left(\frac{dF}{d(x_2, x_3)}, \frac{dF}{d(x_3, x_1)}, \frac{dF}{d(x_1, x_2)}, x_1, x_2, x_3\right) + h = 0, \quad (2)$$

obtained by substituting in H for the p's the derivatives $\frac{dF}{d(x_i, x_s)}$ of a *function of a line of the first degree with respect to the plane* x_i, x_s (cf. Nos. 61 to 65), the same relations hold as those discovered by JACOBI to hold between the canonical equations and his partial differential equation. In the case we are considering, for the canonical equations there have been substituted the partial differential equations (1), and for JACOBI's partial differential equation there has been substituted an equation (2) of a new type, a functional derivative equation.

M. FRÉCHET has discussed this extension of the JACOBI-HAMILTON theory for the general case ([1]).

§ 2.

123. As an example of the first type of problem, we propose, if possible, to determine a *regular* functional $F\left[\overset{b}{\underset{a}{y(t)}}\right]$ which

---

([1]) Cf. FRÉCHET, (19).

V FUNCTIONAL DERIVATIVE EQUATIONS 157

shall have for derivative at the point x a given functional $\Phi\left[y(t); \ x\right]_a^b$; i.e. we propose to solve the equation

$$F'[y(t); \ x] = \Phi[y(t); \ x]. \tag{1}$$

This problem can evidently be considered as the limiting case of the total differential system

$$\varphi_r(y_1, y_2, \ldots y_n) = \frac{\partial f}{\partial y_r} \qquad (r = 1, 2, \ldots n) \tag{2}$$

when the number n of the variables y_i tends to infinity and a continuous variable t (or x) is substituted for the discontinuous index i (or r). Now in the case of the system (2), the necessary and sufficient condition for the existence of the solution is that the relation

$$\frac{\partial \varphi_r}{\partial y_s} = \frac{\partial \varphi_s}{\partial y_r} \tag{3}$$

shall hold for any pair whatever of indices r and s (conditions of integrability); similarly, for the equation (1), remembering that the second derivative of a functional F at the points x and ξ is a *symmetric function* of x and ξ (No. 29), we get the analogous condition (the *condition of integrability of* (1))

$$\Phi'[y(t); \ x, \xi] = \Phi'[y(t); \ \xi, x]; \tag{4}$$

i.e. the functional derivative at the point ξ of the known functional $\Phi[y(t); x]$ must be a symmetric function of x and ξ ([1]).

This theorem can also be enunciated as follows: (4) *expresses the necessary and sufficient condition that the functional* $\Phi[y(t); x]$ *may be considered as the functional derivative at the point x of another functional* $F[y(t)]$; or alternatively, that the expression

$$\int_a^b \Phi[y(t); \ x] \, \delta y(x) \, dx$$

may be the exact total differential or *variation* δF of a functional $F\left[y(t)\right]_a^b$.

---

[1] Cf. VOLTERRA, (87) of Bibliography I, p. 43.

§ 3.

124. In order to prove this assertion we must first establish an extension of STOKES's theorem (which is applicable in the case of a system (2) (No. 123) in three variables y_1, y_2, y_3) to the case of an infinite number of variables. If l denotes the boundary of a surface σ and n the normal to the surface, STOKES's theorem is

$$\int_l (X_1 \, dy_1 + X_2 \, dy_2 + X_3 \, dy_3) = \int_\sigma \left[\left(\frac{\partial X_3}{\partial y_2} - \frac{\partial X_2}{\partial y_3} \right) \cos n y_1 \right.$$
$$\left. + \left(\frac{\partial X_1}{\partial y_3} - \frac{\partial X_3}{\partial y_1} \right) \cos n y_2 + \left(\frac{\partial X_2}{\partial y_1} - \frac{\partial X_1}{\partial y_2} \right) \cos n y_3 \right] d\sigma, \quad (1)$$

or putting $y_i = y_i(u, v)$,

$$\int_s \left(X_1 \frac{dy_1}{ds} + X_2 \frac{dy_2}{ds} + X_3 \frac{dy_3}{ds} \right) ds$$
$$= \int_\omega \left[\left(\frac{\partial X_3}{\partial y_2} - \frac{\partial X_2}{\partial y_3} \right) \frac{d(y_2, y_3)}{d(u, v)} + \cdots \right] du \, dv, \quad (2)$$

where s denotes the boundary of the region ω of the plane u, v on which the points of σ are represented. Extending the formula (2) to the case of n variables $y_1, y_2, \ldots y_n$, we have

$$\int_s \sum_1^n X_i \frac{dy_i}{ds} ds = \frac{1}{2} \int_\omega \sum_1^n{}_r \sum_1^n{}_k \left(\frac{\partial X_r}{\partial y_k} - \frac{\partial X_k}{\partial y_r} \right) \frac{d(y_k, y_r)}{d(u, v)} du \, dv. \quad (3)$$

To pass from n to an infinite number of variables, we shall have to consider, not functions $X_i(y_1, y_2, \ldots y_n)$ depending on the discontinuous index i, but functionals $\Phi \left[y(t); x \right]_a^b$ depending on a continuous variable x, and to replace sums by integrals. We shall then have

$$\int_s ds \int_a^b \Phi \left[y(t, s); x \right]_a^b \frac{\partial y(x, s)}{\partial s} dx$$
$$= \frac{1}{2} \int_\omega du \, dv \int_a^b \int_a^b \left\{ \Phi' \left[y(t; u, v); x, \xi \right]_a^b \right.$$
$$\left. - \Phi' \left[y(t; u, v); \xi, x \right]_a^b \right\} \frac{d[y(\xi; u, v), y(x; u, v)]}{d(u, v)} d\xi \, dx, \quad (4)$$

where $y(t, s)$ denotes the value of the function $y(t; u, v)$ when the point u, v of the region ω coincides with a point s of the boundary, and $\Phi'\,[y(t;u,v);x,\xi]$, etc., denotes the functional derivative at the point ξ of the functional $\Phi\,[y(t;u,v);x]$. With this extension of STOKES's theorem to the case of functionals we see that the integral on the left, taken round the closed curve which forms the boundary of the field ω, is certainly zero if the condition (4) of No. 123 is satisfied. This is equivalent to saying that in this case the integral

$$I = \int_{s_0}^{s_1} ds \int_a^b \Phi\left[\underset{a}{\overset{b}{y(t,s)}};\, x\right] \frac{\partial y(x,s)}{\partial s} dx$$

does not depend on the path followed to pass from $y_0(t) = y(t,s_0)$ to $y_1(t) = y(t,s_1)$; and as its variation δI when we pass from y_1 to $y_1 + \delta y_1$ is

$$\delta I = \int_a^b \Phi\,[y_1(t);\, x]\, \delta y_1\, dx,$$

we see that the functional $F[y(t)]$ that we are in search of, which has $\Phi[y(t);x]$ for derivative, is completely determined when the initial value $F[y_0(t)]$ (an arbitrary constant) is given for a particular function $y_0(t)$. The value of the functional for any other function $y_1(t)$ will be obtained by taking an arbitrary function of two variables $y(t,s)$ such that $y(t,s_0) = y_0(t)$, $y(t,s_1) = y_1(t)$, and putting [1]

$$F[y_1(t)] = F[y_0(t)] + \int_{s_0}^{s_1} ds \int_a^b \Phi\left[\underset{a}{\overset{b}{y(t,s)}};\, x\right] \frac{\partial y(x,s)}{\partial s} dx, \quad (5)$$

an expression which extends to functionals the well-known formula giving the integral of the total differential system (2) of No. 123.

Numerous results relating to functional derivative equations, analogous to those for total differential systems, have been obtained by PAUL LÉVY [2].

§ 4.

125. We shall now consider the functional derivative equation

$$\int_a^b F'\,[y(t);x]\, y(x)\, dx = 0 \qquad (1)$$

---

[1] Cf. VOLTERRA, (87) of Bibliography I, p. 48.
[2] Cf. P. LÉVY. (35)–(42).

in the unknown (regular) functional $F[y(t)]$. It can evidently be considered as obtained, by the usual process of passing from the finite to the infinite (Nos. 1 to 5), from the partial differential equation

$$\sum_{1}^{n} \frac{\partial f(y_1, y_2, \ldots y_n)}{\partial y_r} y_r = 0,$$

the general integral of which, as is well known by EULER's theorem, is given by any homogeneous function whatever of degree 0, i.e. by

$$f(y_1, y_2, \ldots y_n) = \varphi \left(\frac{y_1}{\sum_{1}^{n} y_r}, \frac{y_2}{\sum_{1}^{n} y_r}, \ldots \frac{y_n}{\sum_{1}^{n} y_r} \right),$$

where φ denotes an arbitrary function of n variables.

It could be shown on similar lines ([1]) that if $\Phi\left[y(t)\right]_a^b$ denotes an *arbitrary functional*, the general integral of (1) is given by

$$F\left[y(t)\right]_a^b = \Phi\left[\frac{y(t)}{\int_a^b y(\xi)\, d\xi}\right],$$

i.e. by a functional of $y(t)$ which we can call *homogeneous of degree zero*, since its value is unchanged if we substitute for $y(t)$ another function $\varrho y(t)$ proportional to it. We see therefore that in this case the solution depends on an arbitrary *functional* Φ, and not on a constant, as in the preceding case.

An equation of which (1) is a particular case is the following, which has been discussed by E. FREDA:

$$\int_a^b F'\left[y(t)\, x\right]_a^b y(x)\, dx = r\, F\left[y(t)\right]_a^b \qquad (2)$$

(where r is a real number).

It could be shown that (2) is the characteristic equation of homogeneous functionals of degree r, that is, of the functionals which satisfy the condition:

$$F\left[M \cdot f(t)\right]_a^b = M^r\, F\left[f(t)\right]_a^b.$$

([1]) Cf. VOLTERRA, (67).

V FUNCTIONAL DERIVATIVE EQUATIONS

Hence the general integral of (2) can be given by

$$\Phi\left[\frac{y\left(\overset{b}{\underset{a}{(t)}}\right)}{\int_a^b y(\xi)\,d\xi}\right] \cdot \int_a^b \int_a^b \cdots \int_a^b f^{r_1}(\xi_1) f^{r_2}(\xi_2) \cdots f^{r_s}(\xi_s)\,d\xi_1\,d\xi_2 \cdots d\xi_s,$$

where Φ is an arbitrary functional and $r_1 + r_2 + \cdots + r_s = r$. General integrals of several functional derivative equations, which can be reduced to equation (2), are easily obtained from the preceding integral [1].

§ 5. Relations between Linear Functional Derivative Equations and Integro-Differential Equations.

126. It is well known that if

$$y_i = y_i(x; c_1, c_2, \ldots c_n) \qquad (i = 1, 2, \ldots n) \tag{1}$$

is a system of integrals of the differential system

$$\frac{d y_i}{d x} = f_i(y_1, y_2, \ldots y_n, x) \qquad (i = 1, 2, \ldots n), \tag{2}$$

and if we solve the system (1) for the arbitrary constants c_i in the form

$$c_i = \varphi_i(y_1, y_2, \ldots y_n, x),$$

then the functions φ_i and any function $\varphi(\varphi_1, \varphi_2, \ldots \varphi_n)$ whatever of them will be solutions of the linear partial differential equation of the first order

$$\frac{\partial \varphi}{\partial x} + \sum_1^n f_i(y_1, y_2, \ldots y_n)\frac{\partial \varphi}{\partial y_i} = 0. \tag{3}$$

If we pass from a finite to an infinite number of variables by the usual process, the differential system (2) is transformed into the *integro-differential* equation

$$\frac{\partial y(x, \xi)}{\partial x} = F\left[y(x, \overset{b}{\underset{a}{t}}); x, \xi\right], \tag{4}$$

---

[1] Cf. E. Freda, (20).

studied above in No. 106, which has for solution a function $y(x, \xi)$ depending, not on n arbitrary constants as in (1), but on an arbitrary *function* $\psi(t)$ (e.g. on the initial value $y(x_0, t) = y_0(t)$ taken by y for $x = x_0$), and given by

$$y(x, \xi) = \psi(\xi) + \Phi\left[\psi(t); \; x, \xi\right]_a^b. \tag{5}$$

Considering x in this equation as a constant parameter, we can solve it with respect to the arbitrary function $\psi(t)$, and we shall have

$$\psi(\xi) = y(x, \xi) + \Omega\left[y(x, t); \; x, \xi\right]_a^b.$$

If now we consider the function y as an independent variable (independent also of the parameter x), the expression

$$\psi(\xi) = y(\xi) + \Omega\left[y(t); \; x, \xi\right]_a^b$$

will be a functional of y (corresponding to the functions $\varphi_i(y_1, y_2, \ldots y_n, x)$ obtained from (1)), which, like any other functional $\varLambda$ of $y(t)$ obtained from the preceding one by means of an *arbitrary functional* Θ,

$$\Theta\left[y(\xi) + \Omega[y(t); x, \xi]_a^b\right]_a^b = \varLambda\left[y(t); \; x\right]_a^b,$$

will satisfy the *linear functional derivative equation of the first order*

$$\frac{\partial \varLambda}{\partial x} + \int_a^b \varLambda'[y(t); \; x, \xi] \, \mathrm{F}[y(x, t); \; x, \xi] \, d\xi = 0, \tag{6}$$

where $\varLambda'[y(t); x, \xi]$ denotes the functional derivative of $\varLambda$ at the point ξ. This is the extension of (3) to the case of an infinite number of variables.

Reciprocally, *in order to integrate a linear functional derivative equation of the first order of the type* (6), *we have only to integrate the integro-differential equation* (4) ([1]).

([1]) Cf. VOLTERRA, (68).

§ 6. Equations of Canonical Type.

127. A case of differential systems (2) which is of particular importance for its applications is that of the *canonical equations*

$$\left.\begin{aligned}\frac{dq_i}{dt} &= \frac{\partial H(q_1, q_2, \ldots q_n, p_1, p_2, \ldots p_n, t)}{\partial p_i} \\ \frac{dp_i}{dt} &= -\frac{\partial H(q_1, \ldots q_n, p_1, \ldots p_n, t)}{\partial q_i}\end{aligned}\right\}, \quad (7)$$

to which HAMILTON has reduced the equations of dynamics, and the integration of which has been accomplished by JACOBI by means of the integration of a single partial differential equation. In the case of an infinite number of variables, the corresponding problem is the study of the integro-differential equations which we shall call *equations of canonical type*.

If in fact we take a functional $H\left[\underset{a}{\overset{b}{f(\xi)}}, \underset{a}{\overset{b}{\varphi(\xi)}}; t\right]$ depending on all the values of two arbitrary functions f and φ and on the parameter t, and if we denote by $H'_f[f(\xi), \varphi(\xi); t, x]$, $H'_\varphi [f(\xi), \varphi(\xi); t, x]$, the two derivatives at the point x with respect to f and φ respectively, then the integro-differential equations of canonical type will be given by

$$\left.\begin{aligned}\frac{\partial q(t,x)}{\partial t} &= H'_p\left[\underset{a}{\overset{b}{q(t,\xi)}}, \underset{a}{\overset{b}{p(t,\xi)}}; t, x\right] \\ \frac{\partial p(t,x)}{\partial t} &= -H'_q\left[\underset{a}{\overset{b}{q(t,\xi)}}, \underset{a}{\overset{b}{p(t,\xi)}}; t, x\right]\end{aligned}\right\}. \quad (8)$$

For the integration of these equations it is sufficient to determine a functional $V[f(\xi), \alpha(\xi); t]$ of two functions f and α, satisfying the functional derivative equation

$$\frac{\partial V}{\partial t} + H\left[\underset{a}{\overset{b}{f(\xi)}}, \underset{a}{\overset{b}{V'_f(t,\xi)}}; t\right] = 0,$$

where $V'_f[f(\xi), \alpha(\xi); t, x] = V'_f(t, x)$ denotes the derivative of V with respect to f at the point x. The solutions $p(t, x)$, $q(t, x)$ of the canonical equations (8) are obtained from the solution V of this equation by means of the (integral) equations

$$V'_\alpha\left[q(t,\xi),\alpha(\xi); t,x\right] = b(x),$$
$$V'_q\left[q(t,\xi),\alpha(\xi); t,x\right] = p(t,x),$$

with $b(x)$ an arbitrary function, and certain hypotheses as to the variation $\delta V'_\alpha$ of V'_α [1].

In the same way POISSON's theory of parentheses and STÄCKEL's theory of the separation of the variables can be extended to the case of an infinite number of variables [2].

In a recent paper W. HEISENBERG and W. PAULI [3] discuss a particular case of the equations (8) for quantum mechanics. What they call a *Hamilton* or *functional derivative* is only the first functional derivative of a functional (cf. No. 27) [4]. There is no reason for calling it a HAMILTON derivative.

§ 7. *Functional Derivative Equations of the Second Order.*

128. In addition to the functional derivative equations of the first order which we have so far considered, other interesting types are the *linear functional derivative equations of the second order* of the type

$$\int_a^b \int_a^b F''\left[y(t); \xi,\eta\right] K(\xi,\eta)\,d\xi\,d\eta = 0$$

(considered also by GATEAUX [5] in his attempts to extend the theory of the potential to functional spaces), and

$$\int_a^b \int_a^b F''\left[y(t); \xi,\eta\right] H(\xi,\eta)\,d\xi\,d\eta + \int_a^b F''\left[y(t); \xi,\xi\right] d\xi = 0,$$

equations which correspond to linear partial differential equations of the second order *with constant coefficients* (in particular, LAPLACE's equations), and which we shall call respectively equations of the first and second kind.

---

[1] Cf. VOLTERRA, (68).
[2] Cf. VOLTERRA, (87) of Bibliography I, and (68) below.
[3] Cf. HEISENBERG and PAULI, (81) of Bibliography VI.
[4] Cf. also JORDAN and PAULI jr., (29).
[5] Cf. GATEAUX, (43) of Bibliography I; P. LÉVY, (43).

P. JORDAN and W. PAULI jr.(¹) have used functionals in the new quantum electrodynamics. The theorem of energy in the case of a one-dimensional continuum (not relativistic) leads to the following functional derivative equation analogous to SCHRÖDINGER's partial differential equation:

$$-\left(\frac{h}{4\pi}\right)^2 \int_{x_1}^{x_2} \Psi'' \left[q(x); x, x \right]_{x_1}^{x_2} + c^2 \left[\int_{x_1}^{x_2} \left(\frac{\partial q}{\partial x}\right)^2 dx \right] \Psi \left[q(x) \right]_{x_1}^{x_2} = E \Psi \left[q(x) \right]_{x_1}^{x_2}.$$

In the general relativistic case of electrodynamics in a vacuum, there are four functional derivative equations playing the same rôle as the SCHRÖDINGER differential equation.

§ 8. *A Functional Derivative Equation of Green's Function.*

129. It is well known that if h_B^A is a regular harmonic function of the point A of a plane area bounded by a closed contour C of length s, which takes on the contour C the same values as $\log \frac{1}{r}$ (where r denotes the distance of the point A from the point B within the area considered), then the function

$$g_A^B = g_B^A = \log \frac{1}{r} - h_B^A,$$

which is called *Green's function* relative to the contour, will be harmonic in the interior of the area, but will become infinite like $\log \frac{1}{r}$ at the point B, and will vanish on the contour C. By means of this function g_B^A we can then express any other function u_A which is harmonic and regular in the area considered, when only the values of u on the contour C (and not those of the normal derivative) are known, by the formula

$$u_A = \frac{1}{2\pi} \int_C u_M \frac{dg_M^A}{dn} ds$$

(M variable along C, $\dfrac{dg_M^A}{dn}$ the normal derivative), a formula which solves what is called DIRICHLET's problem.

(¹) Cf. JORDAN and PAULI, (29).

GREEN's function g_B^A will therefore depend, not only on the point A and the pole B where it becomes infinite, but also *on the contour* C; it will therefore be a *function of the line* C (No. 9), or, if the expression is preferred, a functional of the functions which together determine the closed curve C. This particular functional of the line C satisfies a notable total differential equation, found by HADAMARD [1]. If in fact δg_B^A denotes the variation of this functional when we keep A and B fixed (within C) and make the contour C vary (or more precisely, make every point M of C vary by δn along the normal to C), then this variation is given by the formula

$$\delta g_B^A = -\frac{1}{2\pi}\int_C \frac{dg_M^A}{dn}\frac{dg_B^M}{dn}\delta n\,ds, \qquad (1)$$

which is precisely HADAMARD's equation which is satisfied by GREEN's function g_B^A, considered as a functional of the contour C.

Other analogous cases have been treated by P. LÉVY[2]. G. KRALL has treated the case of a multiply-connected region. He has also applied analogous methods to the formulae of SOMIGLIANA and to other interesting questions [3].

An extension of (1) to the theory of elasticity was given by G. ZANONI [4].

M. B. HOSTINSKY[5] shows that HADAMARD's equation above can be applied to GREEN's function of the equation

$$\Delta^2 u + K^2 u = 0.$$

---

[1] Cf. J. HADAMARD, (23).
[2] Cf. P. LÉVY, (39)—(42).
[3] Cf. G. KRALL, (30)—(35).
[4] Cf. G. ZANONI, (69).
[5] Cf. B. HOSTINSKY, (87) of Bibliography VI.

BIBLIOGRAPHY V.

(1) AMOROSO (L.).—Sopra un'equazione integro-differenziale del tipo parabolico. (*R. Acc. dei Lincei. Rend.*, Vol. XXI, Series 5, 2nd half-year, 1912.)

(2) AMOROSO (L.).—Estensione di alcuni precedenti risultati. (*R. Acc. dei Lincei. Rend.*, Vol. XXI, Series 5, 2nd half-year, 1912.)

(3) AMOROSO (L.).—Sopra un nuovo tipo di equazione integro-differenziale. (*Atti Soc. Ital. Prog. Scienze*, 6a riunione, Genoa, 1912.)

(4) ANDREOLI (G.).—Sulle espressioni lineari integro-differenziali. (*R. Acc. dei Lincei. Rend.*, Vol. XXII, Series 5, 2nd half-year, 1913.)

(5) ANDREOLI (G.).—Sulle equazioni integrali miste ed integro-differenziali. (*R. Acc. dei Lincei. Rend.*, Vol. XXIII, Series 5, 1st half-year, 1913.)

(6) ANDREOLI (G.).—Sulle più generali equazioni integrali ed integrodifferenziali ad una variabile. (*R. Acc. dei Lincei. Rend.*, Vol. XXIII, Serie 5, 2nd half-year, 1914.)

(7) ANDREOLI (G.).—Su alcune equazioni integro-differenziali lineari a due e più variabili. (*Atti R. Ist. Veneto*, Vol. 74, 1914-15. Venice, 1915.)

(8) BAERI (L.).—Sulle equazioni integro-differenziali della forma

$$\frac{d^2\varphi(x)}{dx^2} = \mathrm{f}[x, \varphi(x), \varphi'(x), \int_0^x g(x, y), \varphi(y), \varphi'(y) dy].$$

(*Rend. del Circ. Mat. di Palermo*, Vol. 44, 1920.)

(9) BARNET (A.).—Integro-differential equations with constant limits of integration. (*Bull. of American Math. Soc.*, Vol. 26, 1920.)

(10) BARNET (I. C.).—Linear partial differential equations with a continuous infinitude of variables. (*American Journal of Math.*, Vol. 45.)

(11) BEDARIDA (C. H.).—Un problema al contorno in un tipo di equazioni integro - differenziali. (*Rend. del Circ. Mat. di Palermo*, Vol. XLVIII, 1924.)

(12) CRUDELI (U.).—Contributo alla teoria di certe equazioni funzionali. (*R. Acc. dei Lincei. Rend.*, Vol. XVIII, Series 5, 2nd half-year, 1909.)

(13) DANIELE (E.).—Sulle equazioni differenziali e le equazioni integrodifferenziali correlative. (*R. Acc. dei Lincei. Rend.*, Vol. XXVI, Series 5, 1st half-year, 1917.)

(14) EVANS (G. C.).—Sull' equazione integro-differenziale di tipo parabolico. (*R. Acc. dei Lincei. Rend.*, Vol. XXI, Series 5, 2nd half-year, 1912.)

(15) EVANS (G. C.).—Sul calcolo della funzione di Green per le equazioni differenziali ed integro-differenziali di tipo parabolico. (*R. Acc. dei Lincei. Rend.*, Vol. XXII, Series 5, 1st half-year, 1913.)
(16) EVANS (G. C.).—The Cauchy problem for integro-differential equations. (*American Math. Soc. Trans.*, Vol. XV, No. 2, 1914.)
(17) EVANS (G. C.).—Sopra un'equazione integro-differenziale di tipo Bôcher. (*Rend. Seminario Mat. R. Univer. di Roma*, 1920.)
(18) FISCHER (C. A.).—Equations involving the partial derivatives of a function of a surface. (*American Journal of Math.*, Vol. 39, 1917.)
(19) FRÉCHET (M.).— Sur une extension de la méthode de Jacobi-Hamilton. (*Ann. di Mat.*, Series 3, Vol. XI, 1905.)
(20) FREDA (E.).—Il teorema di Eulero per le funzioni di linea omogenee. (*R. Acc. dei Lincei. Rend.*, Vol. XXIV, Series 5, 1915.)
21) FUBINI (G.).—Alcuni nuovi problemi del calcolo delle variazioni con applicazioni alla teoria delle equazioni integro-differenziali. (*Ann. di Mat.*, Series 3, Vol. XX, 1913.)
(22) GRAMEGNA (M.).—Serie di equazioni differenziali lineari ed equazioni integro-differenziali. (*Atti. Acc. di Torino*, Vol. XLV, 1910.)
(23) HADAMARD (J.).—Mémoire sur le problème d'analyse relatif à l'équilibre des plaques élastiques encastrées. (*Mémoires présentés par divers savants à l'Académie des Sciences de l'Institut de France*, Vol. XXXIII, 1908).
(24) HADAMARD (J.).—Sur les ondes liquides. (*R. Acc. dei Lincei. Rend.*, Vol. XXV, Series 5, 1916.)
(25) HADAMARD (J.).—Sur certaines solutions d'une équation aux dérivées fonctionnelles. (*C. R. de l'Acad. des Sciences de Paris*, Vol. 170, 1920.)
(26) HEBRONI (P.).—Über sogenannte zweigliedrige, kontinuisierte Matrizen und ihre Anwendung auf Integral- und Integro-differentialgleichungen. (*Monatsheften für Math. und Phys.*, Vol. XXXIII, Vienna, 1923.)
(27) HILB (E.).—Zur Theorie der linearen Integrodifferentialgleichungen. (*Math. Ann.*, Vol. 77, 1916.)
(28) HU (M. T.).—Linear integro-differential equations with a boundary condition. (*American Math. Soc. Trans.*, Vol. 19, 1918.)
(29) JORDAN (P.) and PAULI (W.) jr. — Zur Quantenelektrodynamik ladungsfreier Felder. (*Zeitschrift für Physik*, Vol. 47, 1928.)
(30) KRALL (G.)—Sulla deformazione infinitesima del campo di integrazione nelle equazioni di Fredholm. (*R. Acc. dei Lincei. Rend.*, Vol. IV, Series 6, 2nd half-year, 1926.)
(31) KRALL (G.).—Sulle funzioni di Green relative a campi pluriconnessi. (*R. Acc. dei Lincei. Rend.*, Vol. V, Series 6, 1st half-year, 1925.)
(32) KRALL (G.).—Variazione infinitesima delle funzioni di Green relative a campi pluriconnessi. (*R. Acc. dei Lincei. Rend.*, Vol. VI, Series 6, 2nd half-year, 1927.)
(33) KRALL (G.).—Dipendenza funzionale dal contorno del tensore di Green-Somigliana per le equazioni dell' elasticità. (*R. Acc. dei Lincei. Rend.*, Vol. VI, Series 6, 2nd half-year, 1927.)
(34) KRALL (G.).—Variazione del campo nelle equazioni del moto elastico. (*R. Acc. dei Lincei. Rend.*, Vol. VI, Series 6, 2nd half-year, 1927.)

(35) KRALL (G.).—Sur la variation du domaine dans le problème de Dirichlet. (*C. R. de l'Acad. des Sciences de Paris*. Vol. 189, 1929.)
(36) LÉVY (P.).—Sur quelques équations définissant des fonctions de ligne. (*C. R. de l'Acad. des Sciences de Paris*, Vol. 151, 1910.)
(37) LÉVY (P.).—Sur l'integrabilité des équations définissant des fonctions de ligne. (*C. R. de l'Acad. des Sciences de Paris*, Vol. 151, 1910.)
(38) LÉVY (P.).—Sur les équations integro-différentielles definissant des fonctions de ligne (thèse). (Gauthier-Villars, Paris, 1911.)
(39) LÉVY (P.).—Sur les équations aux dérivées fonctionnelles et leur application à la Physique mathématique. (*Rend. del Circ. Mat. di Palermo*, Vol. XXXIII, 1912.)
(40) LÉVY (P.).—Sur la fonction de Green ordinaire et la fonction de Green d'ordre deux relative au cylindre de révolution. (*Rend. del Circ. Mat. di Palermo*, Vol. XXXIII, 1912.)
(41) LÉVY (P.).—Sur l'integration des équations aux dérivées fonctionnelles partielles. (*Rend. del Circ. Math. di Palermo*, Vol. XXXVII, 1914.)
(42) LÉVY (P.).—Sur l'allure des fonctions de Green et de Neumann dans le voisinage du contour. (*Acta Math.*, Vol. 42, 1920.)
(43) LÉVY (P.).—Sur la généralisation de l'équation de Laplace dans le domaine fonctionnel. (*C. R. de l'Acad. des Sciences de Paris*, Vol. 168, 1919.)
(44) PÉRÈS (J.).—Résolution des problèmes aux limites relatifs à une équation intégro-différentielle de M. Volterra. (*Rend. del Circ. Mat. di Palermo*, Vol. XXXV, 1913.)
(45) POMEY (L.).—Sur les équations intégro-différentielles linéaires à plusieurs variables. (*C. R. de l'Acad. des Sciences de Paris*, Vol. 177, 2nd half-year, 1923.)
(46) POMEY (L.).—Sur les équations intégro-différentielles. (*Thèse*. Gauthier-Villars, Paris, 1924.)
(47) POMEY (L.).—Sur les singularités des équations différentielles et intégro-différentielles à une ou plusieurs variables. (*Extrait des C. R. de l'Acad. des Sciences de Paris*, Paris, 1924.) (The demonstrations of the theorems enunciated in this paper are given by L. POMEY: Sur les équations intégro-différentielles à plusieurs variables et leurs singularités; *Proceedings of the International Mathematical Congress*, Toronto, 1924.)
(48) POMEY (L.).—Sur les solutions régulières des équations différentielles et intégro-différentielles à une ou plusieurs variables. (*Giornale di Matematiche di Battaglini*, Vol. LXIII (1925), 15 of Series 3.)
(49) POMEY (L.).—Intégration d'un système comprenant une infinité d'équations différentielles ordinaires à une infinité d'inconnues. (*C. R. de l'Acad. des Sciences de Paris*, 1st half-year, 1926.)
(50) POMEY (L.).—Sur les équations intégro-différentielles normales d'ordre infini. (*C. R. de l'Acad. des Sciences de Paris*, 1st half-year, 1927.)
(51) POMEY (L.)—Sur les équations intégro-différentielles aux dérivées partielles d'ordre infini, dont la solution a le même domaine d'existence que les coefficients. (*C. R. de l'Acad. des Sciences de Paris*, 1st half-year, 1927.)

(52) SASSMANNSHAUSEN (A.).—Zur Theorie der linearen Integro-differentialgleichungen. (*Deutsche Math. Ver. XXV, Diss. Giessen*, Berlin, 1916.)

(53) SCHLESINGER (L.)—Sur les équations intégro-différentielles. (*C. R. de l'Acad. des Sciences de Paris*, Vol. 158, 1st half-year, 1914.)

(54) SCHLESINGER (L.).—Zur Theorie der linearen Integro-differentialgleichungen. (*Jahr. d. deutsch. Math. Ver.*, Vol. 24, 1915; reprinted in *M. Tud. Akad. Math. és termesz. Értesitö*, Vol. XXXIV, 1916.)

(55) SCHÖNBAUM (E.).—On an integro-differential equation (in Czech). (*Rozpravy*, 29, No. 15, 1920.)

(56) SINIGALLIA (L.)—Sopra un'equazione integro-differenziale del tipo ellittico. (*R. Acc. dei Lincei. Rend.*, Vol. XXIV, Series 5, 1st half-year, 1915.)

(57) TRICOMI (A.).—Su di una classe di equazioni alle derivate funzionali. (*R. Acc. dei Lincei. Rend.*, Vol. XXX, Series 5, 1921.)

(58) VERGERIO (A.).—Sopra un tipo di equazioni integro-differenziali. (*R. Acc. dei Lincei. Rend.*, Vol. XXVIII, Series 5, 1919.)

(59) VOLTERRA (V.).—Sopra un'estensione della teoria Jacobi-Hamilton del calcolo delle variazioni. (*R. Acc. dei Lincei. Rend.*, Vol. VI, 1890.)

(60) VOLTERRA (V.).—Sulle equazioni integro-differenziali. (*R. Acc. dei Lincei. Rend.*, Vol. XVIII, Series 5, 1st half-year, 1909.)

(61) VOLTERRA (V.).—Osservazioni sulle equazioni integro-differenziali ed integrali. (*R. Acc. dei Lincei. Rend.*, Vol. XIX, Series 5, 1st half-year, 1910.)

(62) VOLTERRA (V.).—Sopra una proprietà generale delle equazioni integrali ed integro-differenziali. (*R. Acc. dei Lincei. Rend.*, Vol. XX, Series 5, 2nd half-year, 1911.)

(63) VOLTERRA (V.).—Equazioni integro-differenziali con limiti costanti. (*R. Acc. dei Lincei. Rend.*, Vol. XX, Series 5, 1st half-year, 1911.)

(64) VOLTERRA (V.).—Sur les équations intégro-différentielles et leurs applications. (*Acta Math.*, Vol. 35, 1912.)

(65) VOLTERRA (V.).—Vibrazioni elastiche nel caso della eredità. (*R. Acc. dei Lincei. Rend.*, Vol. XXI, Series 5, 2nd half-year, 1912.)

(66) VOLTERRA (V.).—Sopra equazioni integro-differenziali aventi limiti costanti. (*R. Acc. dei Lincei. Rend.*, Vol. XXII, Series 5, 2nd half-year, 1913.)

(67) VOLTERRA (V.).—Sulle equazioni alle derivate funzionali. (*R. Acc. dei Lincei. Rend.*, Vol. XXIII, Series 5, 1st half-year, 1914.)

(68) VOLTERRA (V.).—Equazioni integro-differenziali ed equazioni alle derivate funzionali. (*R. Acc. dei Lincei. Rend.*, Vol. XXIII, Series 5, 1st half-year, 1914.)

(69) ZANONI (G.).—Estensione della equazione alle derivate funzionali di Hadamard per le funzioni di Green all'elasticità. (*R. Acc. dei Lincei. Rend.*, Vol. XXXIII, Series 5, 1924.)

(70) ZEILON (N.).— Sulle soluzioni fondamentali delle equazioni integro-differenziali. (*R. Acc. dei Lincei. Rend.*, Vol. XXIV, Series 5a, 1st half-year, 1915.)

(71) GRAFFI (D.).—Sulle funzioni di varietà vettoriali. (*R. Acc. dei Lincei. Rend.*, Vol. VI, Series 6, 2nd half-year, 1927.)
(72) GRAFFI (D.).—Sull' induzione magnetica. (*R. Acc. dei Lincei. Rend.*, Vol. VI, Series 6, 2nd half-year, 1927.)
(73) GRAFFI (D.).—Sui problemi dell' ereditarietà lineare. (*Nuovo Cimento*, New Series, Year V, 1928.)
(74) GRAFFI (D.).—Sulla teoria delle oscillazioni elastiche con ereditarietà. (*Nuovo Cimento*, New Series, Year V, 1928.)
(75) GRAFFI (D.).—Su un metodo di calcolo delle proprietà di corpi prossimi alla sfera e al cilindro. (*Bollettino Unione Matematica Italiana*, 15 Dec. 1928.)

CHAPTER VI

APPLICATIONS. OTHER DIRECTIONS OF THE THEORY OF FUNCTIONALS.

Section I.

§ 1. *Calculus of Variations.*

130. It would be absolutely impossible to speak of *all* the applications of the theory of functionals and the related theories; we shall mention only a few, rather by way of examples. It is only necessary to remember that the *calculus of variations* can be considered as a particular chapter in the theory of functionals to estimate the vast extent of its possible applications.

Without going into particulars about the *calculus of variations* — the more so as it is of earlier date than the creation of functional analysis — we shall merely remind the reader that the methods of discussing many *natural problems can be reduced to problems of the calculus of variations*. For instance, it is well known that by means of HAMILTON's principle the equations of *dynamics* can be brought within the calculus of variations. But HAMILTON's principle can be developed in two different directions: the so-called *principle of stationary action* and the *principle of variable action*; and the latter leads to a new application of the theory of functionals. In it, in fact, the action is considered as a *function* of the final values of the integrals and of the time, and if we consider a system with n degrees of freedom, we find that the action is effectively an ordinary function of $n+1$ variables. If however the system is continuous and with an infinite number of degrees of freedom, we reach a case in which the action must be considered as a function of an infinite number of variables, and therefore as a functional.

It follows that the extension of the principle of variable action to the cases of *electricity, magnetism, elasticity,* and so on, and in general to the classical questions of mathematical physics, leads to a corresponding series of principles which cannot be enunciated without the terminology of functionals, and find their proper development within the sphere of the theory of functionals.

We have insisted on this point from the very beginning of our researches (cf. No. 122). It is pleasant to see that this is now also being confirmed in recent studies on the most modern theories of physics[1].

131. The calculus of variations was also employed to prove some celebrated existence theorems, the best known of which is the so-called DIRICHLET's principle.

Its history is well known and the contributions made to it by GAUSS, THOMSON, DIRICHLET, and RIEMANN are continually referred to. After the objections raised by WEIERSTRASS this method of proof was abandoned (cf. No. 133) until it was taken up again by ARZELÀ[2], who applied it to the principles of the theory of functionals. ARZELÀ's attempt was taken up again by D. HILBERT[3] with complete success and subsequently continued by various others[4].

132. In Chapter I (Nos. 17 and 18) we referred to the new outlook given to the calculus of variations by the effect on it of functional analysis, and we mentioned HADAMARD's masterly work[5], and also TONELLI's treatise[6], which applies the new methods very completely. Everyone knows the example given by SCHWARZ and PEANO[7], that a polyhedron can approach a surface indefinitely without the area of the polyhedron tending towards the area of the surface; this remark is sufficient to indicate the immense revolution that must be made in the fundamental concepts of the calculus of variations. Besides the principles of the functional calculus, TONELLI has made use of the idea of semicontinuity (Nos. 17 and 18), which has a very great value. These are the principal novelties constituting the originality of his treatise.

---

[1] Cf. HEISENBERG and PAULI, (81).
[2] Cf. ARZELÀ, (1) and (2) of Bibliography I.
[3] Cf. HILBERT, (83), (84).
[4] Cf. TONELLI, (78) of Bibliography I, Vol. I, p. 26.
[5] Cf. HADAMARD, (48) of Bibliography I.
[6] Cf. TONELLI, (78) of Bibliography I.
[7] Cf. SCHWARZ, (163), and PEANO, (134).

§ 2. Integral Equations.

133. Passing to *integral equations* we may remark that the development of this chapter of analysis has been so great in the last few years that it is impossible to summarise, not merely its development, but also all the applications that have been made of it([1]). We shall refer to some which cannot possibly be passed over in silence, beginning with the so-called *contour problems (partial differential equations of elliptic type)*.

The typical case is that of LAPLACE's equations. The solutions $\varphi(x, y)$ (harmonic functions), regular within a certain area (in the case of functions of two variables), are completely determined when we know the values taken by them on the contour s of the area in question (cf. No. 131). *Dirichlet's problem* consists in actually constructing them when these values are given on the contour; a well-known method of solving it is to determine GREEN's function relative to the contour (a functional of the lines constituting it; cf. No. 129) and then integrate an expression containing the assigned values and the derivative of GREEN's function taken in the direction of the normal to the contour. But the determination of GREEN's function is often at least as difficult a problem.

If instead we make use of integral equations, we can put the harmonic function φ in the form of a *simple layer potential*

$$\varphi_A = \int_s \varrho(s) \log \frac{1}{r(s, A)} ds, \tag{1}$$

where φ_A denotes the value of φ at the point A of the area, and $r(s, A)$ the distance of A from the point s of the contour, and then try to determine the unknown density $\varrho(s)$ from the integral equation of the first kind

$$\varphi(s_1) = \int_s \varrho(s) \log \frac{1}{r(s\ s_1)} ds, \tag{2}$$

which is obtained from (1) by making the point A coincide with a point s_1 of the contour (so that $r(s, s_1)$ denotes the distance between the two points s, s_1 of the contour); an equation of the first kind which also has its kernel infinite for $s = s_1$.

([1]) For a survey of the whole question, together with an extensive bibliography, cf. H. T. DAVIS, (11) of Bibliography II.

VI APPLICATIONS 175

To avoid the difficulties of this equation, we can, following NEUMANN and POINCARÉ, put the harmonic function φ in the form of a *double layer potential*

$$\varphi_A = \int_s \varrho(s) \frac{d \log \frac{1}{r(s,A)}}{dn} ds \qquad (3)$$

and then try to determine the unknown density $\varrho(s)$ from the integral equation obtained by making the point A coincide with a point s_1 of the contour[1],

$$\varphi(s_1) = \pi \varrho(s_1) - \int_s \varrho(s) \frac{d \log \frac{1}{r(s,s_1)}}{dn} ds, \qquad (4)$$

an equation of the second kind with a finite kernel

$$F(s, s_1) = \frac{1}{r(s, s_1)} \cos nr,$$

for which it is easy to show that the determinant $D_F\left(\frac{1}{\pi}\right)$ (cf. No. 45) is different from zero, and that therefore it has always one and only one solution, which is completely determined by known formulae (cf. No. 45).

§ 3.

134. When FREDHOLM published his first classical memoir[2], in which he took DIRICHLET's problem as his point of departure, VOLTERRA had already published[3] the results of his work on integral equations with variable limits, making the suggestion that integral equations might be considered as the limiting case of ordinary systems of linear equations, and in a brief note[4] in which he discussed the problem of "seiches" (cf. Nos. 132 and 134) he had already solved the problem by using infinite determinants.

HILBERT[5] has since applied FREDHOLM's equations to several problems of analysis, geometry, and physics, and under

---

[1] Cf. FREDHOLM, (71).
[2] Ibid.
[3] Cf. VOLTERRA, (76) of Bibliography II.
[4] Cf. VOLTERRA, (180).
[5] Cf. HILBERT, (23) of Bibliography II.

the impulse of SCHMIDT[1] the theory of expansions in series and of orthogonal functions has made wonderful progress. GOURSAT has given a new method of solving FREDHOLM's equation, and LALESCO[2] has worked out some important singular cases. Researches on VOLTERRA's *integro-functional equations* have recently been undertaken by C. POPOVICI, J. D. TAMARKIN, and C. R. ADAMS[3].

135. In HILBERT's work on integral equations just cited, there is developed a theory of functions of *an infinite number of variables* $x_0, x_1, \ldots x_n, \ldots$, in particular of *linear forms* $\sum_n^\infty a_n x_n$ and *quadratic forms* $\sum_r^\infty \sum_s^\infty a_{rs} x_r x_s$. This theory has proved useful for functions whose squares are summable, which can be defined by means of FOURIER's generalised coefficients a_n of a suitable expansion in series of orthogonal and normal functions. For HILBERT's theory the consideration of "limited" functions is therefore fundamental, i.e. functions such that they always remain limited, whatever $x_0, x_1, \ldots x_n, \ldots$ may be, provided that $\sum_1^\infty x_n^2 \leq 1$. In particular, a linear form $\sum_1^\infty a_n x_n$ is limited only if the series $\sum_1^\infty a_n^2$ is convergent.

VITALI[4] has recently developed a "geometry" in which functions (or points) are in fact represented by a system of "coordinates" a_n, infinite in number, for which the series $\sum_1^\infty a_n^2$ (the square of the "distance" from the origin) converges; among other applications a "generalised absolute differential calculus", which includes RICCI's classical calculus, is developed in this way.

HILBERT's theory of functions of an infinite number of variables has recently been invoked in justification of the expansions relating to matrices which are of interest for the theory of *quanta*. In this case, however, HILBERT's hypotheses about the "limitation" of these functions of an infinite number of variables usually fail[5].

---

[1] Cf. SCHMIDT, (60) of Bibliography II.
[2] Cf. LALESCO, (30) of Bibliography II.
[3] Cf. POPOVICI, (147); TAMARKIN, (167); ADAMS, (1).
[4] Cf. VITALI, (176).
[5] Cf. JORDAN and PAULI, (29) of Bibliography V.

§ 4. Seiches.

136. An interesting application of integral equations is the study of the *oscillations or tides of lakes*. The phenomenon was first observed, under the name of *seiches*, in the Lake of Geneva, where, owing to its elongated form, the changes of level sometimes reach a couple of metres; it was for a long time considered as a curiosity, until a serious study of the phenomenon was made by MÉRIAN([1]), FOREL([2]), and others.

We shall consider first the particular case of a heavy liquid, bounded by a free surface and by rigid walls, which is making oscillations parallel to a vertical plane x, y (where y is vertical) and independent of the third coordinate z; and we shall try to find the periods of the oscillations. If $\varphi(x, y, t)$ denotes the velocity potential, we shall have within the liquid

$$\Delta^2 \varphi = 0,$$

on the free surface

$$\frac{\partial^2 \varphi}{\partial t^2} = \alpha \frac{\partial \varphi}{\partial n},$$

and on the rigid walls $\dfrac{\partial \varphi}{\partial n} = 0$ (where n is the normal to the boundary, α a constant, and t the time).

We shall now try to find those solutions φ of our problem which can be put in the form

$$\varphi = \sin ht \, \psi(x, y),$$

i.e. we shall apply the usual method of the separation of the variables. The foregoing equations become respectively

$$\Delta^2 \psi = 0, \quad -h^2 \psi = \alpha \frac{\partial \psi}{\partial n}, \quad \frac{\partial \psi}{\partial n} = 0,$$

and consequently the value at the point A of the harmonic function ψ_A is given by the formula

$$\psi_A - \psi_0 = \int_l f_A(s) \frac{\partial \psi}{\partial n} ds,$$

([1]) Cf. MÉRIAN, (119).
([2]) Cf. FOREL, (53); with a bibliography.

where $f_A(s)$ is a kernel[1] obtained from the known solution of the NEUMANN problem, and l is the portion of the x-axis included within the free horizontal surface of the liquid in the state of equilibrium. We shall then have, by the foregoing equations,

$$\psi_A - \psi_0 = -\frac{h^2}{\alpha} \int_l f_A(x)\psi(x)\,dx,$$

and for $\psi_0 = 0$, and A at the point x_1 of the boundary,

$$\psi(x_1) + \frac{h^2}{\alpha} \int_l f(x,x_1)\psi(x)\,dx = 0. \tag{1}$$

We thus see that the values $\psi(x)$ taken by the harmonic function ψ on the boundary (values which completely determine it) are the values of a function which is a solution of the homogeneous integral equation (1), which, however, has its kernel infinite for $x = x_1$; but by applying the process of iteration we can reduce it to another homogeneous integral equation which has its kernel finite. In order that $\psi(x)$ may not be identically zero, the determinant function $D(\lambda)$ of this integral equation will therefore have to vanish (cf. No. 45); and substituting for λ its expression, which in the case we are considering is a function of h and α, we get a transcendental equation which gives the possible values of h, and therefore the possible periods of the oscillations of the liquid.

§ 5. *Vibrations of Membranes.*

137. Another problem which is connected with the preceding one is that of the *vibrations of membranes*, the equation of which is

$$\frac{\partial^2 u}{\partial t^2} = \frac{\partial^2 u}{\partial x^2} + \frac{\partial^2 u}{\partial y^2}, \tag{1}$$

or, more shortly,

$$\frac{\partial^2 u}{\partial t^2} = \Delta^2 u.$$

This equation can also be treated by the method of the separation of the variables. Putting

$$u = f(t)\,v(x,y),$$

---

[1] Cf. VOLTERRA, (180) below, and (82) of Bibliography II, p. 133.

VI APPLICATIONS 179

we get from (1)

$$\frac{d^2f}{dt^2} + h^2 f = 0,\qquad (2)$$

$$\Delta^2 v + h^2 v = 0,\qquad (3)$$

(h a constant), and therefore $f = \sin(ht+c)$, $u = \sin(ht+c)v(x,y)$. For the value v_A at a point A of the function $v(x,y)$ which satisfies equation (3), we have, by GREEN's theorem (cf. No. 129),

$$v_A = \int_s v(s)\,F_A(s)\,ds - \frac{1}{2\pi}\int_\sigma \Delta^2 v\, g_P^A\, d\sigma,$$

where the first integral is taken over the contour s of the field σ occupied by the membrane in the state of equilibrium; and if we suppose that the membrane is fixed at its boundary, i.e. $v(s) = 0$, we shall have, by (3),

$$v_A = \frac{h^2}{2\pi}\int_\sigma g_P^A\, v\, d\sigma,\qquad (4)$$

g_P^A being the Green's function with pole at A in Dirichlet's problem for s and σ. The function $v(x,y)$ must therefore be a solution of the homogeneous integral equation (4) which has its kernel infinite as $\log\left(\dfrac{1}{r}\right)$ when the variable point coincides with A ($r=0$). By a process of iteration, however, similar to that employed in the previous case of the oscillations of lakes, we can reduce it to another homogeneous integral equation with a finite kernel; and equating to zero the determinant function of this equation (in order to prevent v from being identically zero), we get in this case also a transcendental equation in h which gives the periods of vibration of the membrane.

§ 6.

138. Returning to the problem of the *seiches*, or oscillations of lakes, and not wishing to limit the problem to the case of plane oscillations, we shall refer first to the more recent method of PROUDMAN[1]. Extending a method used by LAGRANGE for the theory of the vibrations of strings, he divides the lake into a number of narrow strips and studies their vibrations, making

---

[1] Cf. PROUDMAN, (149).

them become infinitely narrow and passing to the limit; in his own work, however, he uses infinite determinants. Since then L. MATTEUZZI(^{1}) has pointed out that this method represents precisely the same process of passing to the limit as is found in general in the theory of integral equations; he does not, however, consider it necessary to repeat the general process in this particular case, but applies the general result directly — a proceeding which offers some analogy with the process of passing to the limit which is often used in elementary courses for calculating centres of gravity and moments of inertia, and the successive simplifications acquired as the general methods of the integral calculus become available. He has thus found that the periods of vibration depend on a transcendental equation determined directly by a certain integral equation.

Similar processes can also be applied to the problem of *oceanic tides*; this is one of the methods used by POINCARÉ(^{2}), and by others after him, for the study of this phenomenon.

§ 7. *Some Researches of G. C. Evans and of his School.*

139. Another direction for the application of functions of lines may be found in the simplification of theories of partial differential equations. In fact, additive functions of lines rather than additive functions of point sets present themselves as limits of integrals involving a parameter, when these are extended over regions bounded by curves; the limiting values correspond to the curves rather than to any of the infinite number of point sets which may be associated with a curve. With this kind of application in view, MARIA investigated in his thesis the properties of completely additive functions of plurisegments in the plane, the idea of discard for such functions of curves, the set of singular points, and so forth(^{3}); this study is a natural generalization in a new direction of VITALI's fundamental memoir, which is devoted to such questions for functions $f(x)$ of a single variable (^{4}).

EVANS and BRAY have applied and developed these ideas

(^{1}) Cf. L. MATTEUZZI, (118).
(^{2}) Cf. H. POINCARÉ, (145).
(^{3}) Cf. A. J. MARIA, (117).
(^{4}) Cf. G. VITALI, (175).

in the theory of integro-differential equations of BÔCHER type, the point of departure being BÔCHER's theorem that if u, $\dfrac{\partial u}{\partial x}$, $\dfrac{\partial u}{\partial y}$ are continuous in a region, the equation

$$\int \frac{\partial u}{\partial n} ds = 0$$

holding for all circles in the region implies that the derivatives of all orders exist and are continuous and that u satisfies LAPLACE's equation. EVANS has shown that except for removable discontinuities the same fact is implied by apparently very much less restrictive conditions([1]).

Using this theorem EVANS has been able to solve the discontinuous boundary value problems, of the nature of the DIRICHLET problem, for POISSON's equation when that equation refers to the most general possible distribution of finite positive and negative mass throughout a general simply-connected plane region([2]). The mass function is an arbitrary completely additive function of point sets $\Phi(e)$; corresponding to it there is a single additive function of closed curves $\Phi(s)$, with regular discontinuities, such that for every s where $\Phi(s)$ is continuous, $\Phi(s) = \Phi(e)$, e being the set of points interior to s; the equation considered is then

$$\int_s \frac{\partial u}{\partial n} ds = \Phi(s).$$

The difference of two solutions of this equation, for the class of functions considered, becomes harmonic by the generalization of BÔCHER's theorem. Accordingly it is sufficient to obtain the properties of a principal solution

$$U(M) = \int g(M, P) d\Phi(e),$$

where $g(M, P)$ is the GREEN's function for the region.

140. BRAY and EVANS have shown that the study of a function which is harmonic inside a sphere and can be written as the difference of two not negative functions, harmonic inside the sphere, reduces by means of a generalization of POISSON's

---

([1]) Cf. G. C. EVANS, (26), (31); G. BOULIGAND, (15).
([2]) Cf. G. C. EVANS, (32).

integral to a study of additive functions of lines on the surface of the sphere and STIELTJES integrals based on them ([1]). The same method applies to a discussion of the NEUMANN problem for the sphere ([2]). EVANS and MILES have discussed similarly the DIRICHLET and NEUMANN problems with reference to regions bounded by sufficiently smooth surfaces, and potentials of single and double layers in terms of general distributions of mass ([3]). Here again the appropriate boundary value problems are stated in terms of additive functions of curves on the surface; for example, the limit of the flux through a closed curve as that curve comes up to the surface is such a quantity. These problems are solved by means of integral equations in STIELTJES integrals.

Section II.

§ 1. *Permutable Functions; Functional Groups.*

141. On the theory of permutable functions reference should be made to the writings of EVANS([4]), who was the first to try to construct an algebra of permutable functions of the first kind. He gets over the difficulty of the absence of a zero power by a special and elegant device.

PÉRÈS is the author of a chapter in the theory of permutable functions of the first kind which is fundamental, namely, that of the transformations which maintain the property of composition. We have already spoken of this subject (cf. No. 90), showing how new and rigorous proofs can easily be given of the two fundamental theorems: that several functions which are permutable with another function of the first order are permutable with each other, and that the power by composition with exponent z of a function of the first order is an integral function of z (cf. Nos. 91 and 92).

§ 2.

142. In the chapter on the theory of permutable functions we spoke of integral addition theorems (cf. No. 83), and pointed

---

[1] Cf. G. C. EVANS, (30).
[2] Cf. H. E. BRAY and G. C. EVANS, (17).
[3] Cf. G. C. EVANS and E. R. C. MILES, (33).
[4] Cf. EVANS, (7), (8), and (9) of Bibliography IV.

out, in connection with these theorems, some allied results of great importance due to HADAMARD, who has frequently, and also recently, returned to them([1]). He has observed that the equations of mathematical physics lead to integral addition theorems. To use HADAMARD's own words, "the very definition of a solution in terms of the initial conditions which determine it shows it to us as a *transform* of the system of functions expressed by these conditions; for example, in the case of CAUCHY's problem, the system depending on the functions f and g to which u and $\dfrac{\partial u}{\partial t}$ reduce for $t = t_0$ ([2])."

If now we consider the values f_1 and g_1 taken by u and $\dfrac{\partial u}{\partial t}$ for $t = t_0 + h$, these are obtained from the former values by a *transformation* T_h which depends on the parameter h. Now HUYGENS' principle, at least in one of its senses, expresses the fact that the family of transformations so obtained forms a group, or in other words that

$$T_h \cdot T_k = T_{h+k}.$$

If we take account of the process by means of which the equation considered can be integrated, and which therefore enables us to calculate the transformation T_h, we introduce the *elementary solution* of the equation as an essential element. HUYGENS' principle, i.e. the existence of the group referred to above, leads to an *integral addition theorem*, which is satisfied by the said elementary solution. Further integral addition theorems can be obtained in this way.

§ 3.

143. With a view to their applications in the theory of partial differential equations, and therefore also in the physics of continuous systems, attention had already been directed to the study of continuous groups of *linear functional transformations with one parameter* by KOWALEWSKI([3]), VESSIOT([4]), and more

---

([1]) Cf. HADAMARD, (12) of Bibliography IV.
([2]) Cf. HADAMARD, (79).
([3]) Cf. KOWALEWSKI, (96), (97).
([4]) Cf. VESSIOT, (173), (174).

recently MICHAL[1]. The transformations (which are always reversible) most generally studied are those of VOLTERRA's type

$$y_1(x) = y(x) + \int_0^x K(x,s;a) y(s) ds. \qquad (1)$$

By means of them we pass, for every value of the parameter a, from the function $y(x)$ to the transformed function $y_1(x)$; they have properties exactly similar to those of LIE's continuous transformation groups, defined in a space of n dimensions. The transformations (1) can also be considered as obtained from a LIE group by considering not spaces with a finite number n of dimensions, but functional spaces with a continuous infinity of dimensions. The functionals $F\left[\overset{1}{\underset{0}{y(\tau)}}, \overset{1}{\underset{0}{y'(\tau)}}\right]$ which are *invariant* for these groups of the type (1) have also been studied by MICHAL in his interesting papers; so have groups of transformations with the limits of integration constant (FREDHOLM's type), and analogous invariants. The general expression for linear analytic functional groups with one parameter was found by L. FANTAPPIÈ[2]; T. LEVI-CIVITA[3] had already considered differential equations which are invariant for an extended class of infinite functional groups.

144. Transformations of the general group by HUYGENS' principle considered above are not continuous of order zero, as are those considered by KOWALEWSKI and VESSIOT. They bring in derivatives of the argument-function, which may lead to substantial and far-reaching changes in the results[4].

Section III.

§ 1. *Functional Dynamics.*

145. In a note by VOLTERRA published in 1914[5] the idea was suggested of applying functional analysis to the study of continuous systems, and especially those consisting of swarms of corpuscles not connected by links that can be expressed by

---

[1] Cf. MICHAL, (120) — (124).
[2] Cf. FANTAPPIÈ, (39).
[3] Cf. LEVI-CIVITA, (105) — (107).
[4] Cf. HADAMARD, (79).
[5] Cf. VOLTERRA, (88) of Bibliography I.

means of differential relations. This idea has been exploited in a remarkable recent thesis by G. C. Moisil([1]); he attaches a set of functions, which can be considered as coordinates, to each configuration of a continuous and variable system, which is subject to links of any kind whatever, and so arrives at integrodifferential equations analogous to LAGRANGE's equations. The geometrical scheme of a Riemannian manifold representing the dynamics of the system corresponds to a functional manifold (space of an infinite number of dimensions) having a suitable metric to which the tensor calculus can be applied([2]). In the kinematic part of the question MOISIL studies the functional groups with an infinite number of parameters generated by the displacements of a continuous system; he works on parallel lines to the classical theory of finite groups and reaches analogous results([3]).

The practical applications relate to the movement of a flexible and inextensible wire, but the results obtained are also of interest from the point of view of functional geometry, and especially from the point of view of infinite groups studied by functional methods.

§ 2. *Functional Rotations.*

146. We cannot leave the kinematic questions of functional space and the theory of functional groups without speaking of T. DELSARTE's recent researches([4]).

In the continuous group of linear transformations of n-dimensional space a certain number of sub-groups can be distinguished. Such, for example, are the group of orthogonal transformations and the unimodular group. A simple method of constructing them consists in taking their invariant as given *a priori*, namely, the distance for the one, and the element of volume for the other.

If now we consider the functional space of functions whose squares are summable, and if we take in this space the FREDHOLM group, or group of linear functional transformations

---

([1]) Cf. MOISIL, (127).
([2]) Cf. MOISIL, (125).
([3]) Cf. MOISIL, (126).
([4]) Cf. DELSARTE, (23), (24).

having the special form

$$g(s) = f(s) + \int_0^1 K(s, u) f(u) du = K[f], \tag{1}$$

we can similarly define one of its sub-groups by taking the invariant of this sub-group as given. In particular, the sub-group of "functional rotations" is obtained by imposing on K [f] the property of making the Euclidean distance

$$\int_0^1 f^2(s) ds$$

invariant.
Writing

$$\int_0^1 f^2(s) ds = \int_0^1 g^2(s) ds,$$

we see that the kernel of rotation K (s, u) must satisfy the conditions:

$$R[K] \equiv K(s, t) + K(t, s) + \int_0^1 K(s, u) K(t, u) du$$

$$= S[K] \equiv K(s, t) + K(t, s) + \int_0^1 K(u, s) K(u, t) du = 0,$$

which are in fact equivalent. It follows from these conditions that:
(1) the inverse of the transformation is

$$f(s) = g(s) + \int_0^1 K(u, s) g(u) du;$$

(2) every kernel of rotation K (s, t) is the value taken by the solvent of any skew-symmetric kernel whatever, for $\lambda = \tfrac{1}{2}$;
(3) the functional rotations are displacements in the functional space considered;
(4) every FREDHOLM transformation is the product of a functional rotation by a functional dilatation[1] (we apply this term to a FREDHOLM transformation with a symmetric kernel).

It is further to be remarked that these transformations also leave invariant the notions of volume and surface defined

---

[1] Cf. DELSARTE, (22).

by GÂTEAUX. This is due to the fact that these transformations maintain the normal density of sequences of coordinate functions.

From the geometrical point of view, these transformations have properties very closely resembling those of the rotations of spaces with a finite number of dimensions. It can be shown that they leave invariant a system of doubly extended linear point-sets which are completely orthogonal in pairs, each of them turning on itself round the origin through an angle which is connected with the corresponding singular value of the kernel. In the case where the kernel has only a finite number $2n$ of singular values, we thus have a transformation which is topologically identical with a rotation in space of n dimensions.

§ 3. *The Strain-Energy Functional in the Theory of Elasticity.*

147. L. DONATI [1], in the year 1888, gave an interesting theorem in the theory of elasticity, using the operations of the differentiation and derivation of functionals, which VOLTERRA had introduced a little time before.

DONATI observed that the work of the strain of an elastic body can be considered as a functional of the components u, v, w of the displacements of the different points, or as a functional of the components X, Y, Z of the intensity of the field of force at any point. But the components of the displacements are not only functions of the points of the body, but can also be considered as functionals of the components of the intensity of the field of force. In an analogous way these latter components are not only functions of the points of the body, but can also be considered as functionals of the components of the displacements.

Now the functional derivatives of the work with respect to the functions u, v, w are X, Y, Z, and the functional derivatives of the work with respect to the functions X, Y, Z are u, v, w.

This is DONATI'S theorem for elastic bodies, which is an extension of CASTIGLIANO's theorem [2] for elastic frames.

---

[1] Cf. L. DONATI, (200). *Memorie e note scientifiche*; Bologna, 1925.
[2] Cf. A. CASTIGLIANO: *Théorie de l'équilibre des systèmes élastiques et ses applications.* Turin, 1879.

188   THEORY OF FUNCTIONALS   CHAP.

Section IV.

§ 1. *General Laws of Heredity.*

148. In the previous chapter we spoke of *hereditary* phenomena in order to give a particular example of how integro-differential equations come into existence (cf. No. 115). But the general theory constitutes an application of the general theory of functionals on which we must say a few words.

We have seen how on the hypothesis of heredity there appear in various physical theories terms of the type $z = \mathrm{F}\left[\underset{-\infty}{\overset{t}{f(\tau)}}\right]$, *functionals* depending on the values taken by one or more functions of the time, $f(\tau)$, in the interval between $-\infty$ and the moment in question t; or, in the case of linear heredity, of the type

$$z = \int_{-\infty}^{t} \varphi(t,\tau) f(\tau) \, d\tau.$$

These quantities z, functionals of $f(\tau)$, are also functions in the ordinary sense of the variable t. In many cases, however, as we shall see, by making some simple hypotheses we can particularise the nature of these functionals. A first extremely natural postulate is to suppose that the influence of the heredity corresponding to states a long time before the given moment gradually fades out; more precisely, that, $f(\tau)$ being always limited in absolute value $(<M)$, *the modulus of the variation of the quantity* $z = \mathrm{F}\left[\underset{-\infty}{\overset{t}{f(\tau)}}\right]$ *when $f(\tau)$ varies in any way (its modulus being always $<M$) in the interval $(-\infty, t_1)$* (with $t_1 < t$), *can be made as small as we please by taking the interval (t_1, t) sufficiently large.* This postulate represents the principle of the *dissipation of hereditary action.*

Putting

$$z' = \mathrm{F}\left[\underset{-\infty}{\overset{t+h}{f(\tau-h)}}\right],$$

it may happen that always $z = z'$, whatever f and h may be. This occurs when the phenomenon, while hereditary, always follows the same course if it passes successively through the same conditions, whatever may be the relative position of these conditions in time. If we accept this law as a postulate([1]), at least for isolated systems, we shall say that the heredity is *invariable*, and if the function $y = f(\tau)$ is periodic with period T, then the function $z(t) = F\left[f(\tau)\right]_{-\infty}^{t}$ is periodic with the same period T. If now we represent the phenomenon on a plane by means of a point with cartesian coordinates $y = f(t)$, $z = z(t)$, we get that as t varies the point describes a *closed* curve in every period T. Vice versa it can be shown that if $z(t)$ is periodic whenever $f(t)$ is periodic, and with the same period T for both (whatever T may be), then the heredity is invariable([2]). In view of the method of representing the phenomenon by means of *closed* curves, this principle is also called the principle of the *closed cycle*.

In the case of the closed cycle, which is thus completely equivalent to the case where the heredity is invariable, it can be shown([3]) that all the kernels of the integral equations obtained are simply *functions of the difference* $t - \tau$ of the two variables, i.e. are *permutable of the first kind with unity and with one another* (group of the closed cycle; cf. Chapter IV, No. 87). This result is extremely useful, since we can at once apply the theory of composition which we have studied in an earlier chapter.

Thus, for example, in the case of *elastic torsion*, already considered in No. 115, for which we have supposed the heredity *linear*, we shall have, if the hereditary action is invariable (the case of the closed cycle), the simple integral equation

$$m(t) = h\omega(t) + \int_0^t f(t-\tau)\omega(\tau)\,d\tau$$

with a kernel which is a function of the difference $t - \tau$, instead of the more general equation (5) of No. 115.

Let us now take the case of linear heredity. Let ϱ be a parameter whose value at the time t depends on the parameter q by a relation of linear heredity; i.e. let

$$\varrho(t) = aq(t) + \int_{-\infty}^{t} q(\tau)\Phi(t-\tau)\,d\tau. \tag{1}$$

([1]) Cf. STRANEO, (166).
([2]) Cf. VOLTERRA, (186).
([3]) *Ibid.*, and (86) of Bibliography I.

The function $q(\tau)$ determines the *history* of the parameter q (the primary history), and the function $\varrho(t)$ determines the *history* of the parameter ϱ (the hereditary history). In this case we shall say that the heredity is *complete*. If $\varPhi$ vanishes for values of the argument equal to or greater than T_0, as we have already seen, then the heredity is *limited to the period* T_0, and we have

$$\varrho(t) = aq(t) = \int_{t-T_0}^{t} q(\tau)\varPhi(t-\tau)\,d\tau. \qquad (2)$$

But if the heredity prior to an instant t_0 is negligible, equation (1) will take the form

$$\varrho(t) = aq(t) + \int_{t_0}^{t} q(\tau)\varPhi(t-\tau)\,d\tau. \qquad (3)$$

In this case we shall say that the heredity, is *subsequent* to the instant t_0; the foregoing integral equation can be inverted and we shall have

$$q(t) = \frac{1}{a}\varrho(t) + \int_{t_0}^{t} \varrho(\tau)\,\varphi(t-\tau)\,d\tau,$$

where the function φ is the *solvent kernel* of equation (3). We thus see that, just as the *primary history subsequent to t_0 determines the hereditary history, so the hereditary history subsequent to t_0 determines the primary history*.

The heredity subsequent to t_0 can therefore be completely inverted, and q can be considered as hereditarily dependent on ϱ.

Analogous properties do not hold if the heredity is complete or limited; for example, in these cases various primary histories may correspond to a given hereditary history, unless some restrictive conditions are imposed[1].

If the primary heredity is subsequent to t_0, this does not exclude the possibility of its also being limited to a certain duration T_0; the inverted heredity, however, is not necessarily either limited to the duration T_0, or limited in any way.

---

[1] Cf. VOLTERRA, (190), (188); KOSTITZIN, (92).

VI   APPLICATIONS   191

§ 2. *Hereditary Dynamics.*

149. The general equations of dynamics (LAGRANGE's form) are

$$\frac{d}{dt}\frac{\partial T}{\partial q'_i} - \frac{\partial (T-\Omega)}{\partial q_i} = Q_i, \tag{1}$$

where $q_1, q_2, \ldots q_n$ are the independent coordinates, T the vis viva, $-\Omega$ the potential, $Q_1, Q_2, \ldots Q_n$ the external forces.
If in

$$T = \tfrac{1}{2} \sum_i \sum_s a_{is} q'_i q'_s$$

and in

$$\Omega = \tfrac{1}{2} \sum_i \sum_s b_{is} q_i q_s \tag{2}$$

the coefficients a_{is} and b_{is} can be supposed constant, then the equations take the linear form

$$\sum_s a_{is} q''_s + \sum_s b_{is} q_s = Q_i,$$

from which (if the form (2) is positive) we deduce the theories of small movements.

If linear hereditary effects exist the above equations will be modified and will take the following form[1]:

$$\sum_s a_{is} q''_s + \sum_s b_{is} q_s + \sum_s \int_{-\infty}^{t} \Phi_{is}(t,\tau) q_s(\tau) d\tau = Q_i,$$

and, in the case of the closed cycle and a duration T_0 of the heredity (No. 148),

$$\sum_s a_{is} q''_s + \sum_s b_{is} q_s + \sum_s \int_{t-T_0}^{t} \Phi_{is}(t-\tau) q_s(\tau) d\tau = Q_i,$$

or alternatively

$$\sum_s a_{is} q''_s + \sum_s b_{is} q_s + \sum_s \int_0^{T_0} \Phi_{is}(\tau) q_s(t-\tau) d\tau = Q_i.$$

150. If the number of parameters is 1, i.e. if the system has only one degree of freedom, the preceding equation takes the form $(b > 0)$

$$q'' + bq + \int_0^{T_0} \Phi(\tau) q(t-\tau) d\tau = Q. \tag{3}$$

---

[1] Cf. VOLTERRA, (188), (189).

It can be seen that $\Phi(\tau)$ is a finite continuous negative increasing function which vanishes for $\tau \geqq T_0$ and which satisfies the condition

$$b + \int_0^{T_0} \Phi(\tau) d\tau = m > 0.$$

Equation (3) can also be written

$$q'' + mq - \int_0^{T_0} \Phi(\tau) [q(t) - q(t-\tau)] d\tau = Q.$$

If we put

$$E_p = \tfrac{1}{2} m q^2 - \tfrac{1}{2} \int_0^{T_0} \Phi(\tau) [q(t) - q(t-\tau)]^2 d\tau, \qquad (4)$$

the negative functional $-E_p$ can be called the potential of all the internal forces *both hereditary and non-hereditary*.

§ 3. Hereditary Elasticity.

151. Let $\gamma_{11}, \gamma_{22}, \gamma_{33}, \gamma_{23}, \gamma_{31}, \gamma_{12}$ denote the six elements which determine the strain (or deformation), so that if u, v, w denote the components of the displacement of a point,

$$\left.\begin{aligned} \gamma_{11} &= \frac{\partial u}{\partial x}, \quad \gamma_{22} = \frac{\partial v}{\partial y}, \quad \gamma_{33} = \frac{\partial w}{\partial z} \\ \gamma_{23} &= \frac{\partial v}{\partial z} + \frac{\partial w}{\partial y}, \quad \gamma_{31} = \frac{\partial w}{\partial x} + \frac{\partial u}{\partial z}, \quad \gamma_{12} = \frac{\partial u}{\partial y} + \frac{\partial v}{\partial x} \end{aligned}\right\}; \quad (1)$$

and let t_{ik} (i, $k = 1, 2, 3$) denote the six components (with $t_{ik} = t_{ki}$) along the three axes of the stress (or tension) at each face of an infinitesimal parallelopiped with its edges parallel to the axes. By Cauchy's laws the equations of equilibrium of the elastic body are

$$\left.\begin{aligned} \frac{\partial t_{11}}{\partial x} + \frac{\partial t_{12}}{\partial y} + \frac{\partial t_{13}}{\partial z} &= \varrho X \\ \frac{\partial t_{21}}{\partial x} + \frac{\partial t_{22}}{\partial y} + \frac{\partial t_{23}}{\partial z} &= \varrho Y \\ \frac{\partial t_{31}}{\partial x} + \frac{\partial t_{32}}{\partial y} + \frac{\partial t_{33}}{\partial z} &= \varrho Z \end{aligned}\right\}, \qquad (2)$$

where ρ denotes the density and X, Y, Z the components of the force per unit mass, with the boundary conditions

$$\left. \begin{array}{l} t_{11}\cos nx + t_{12}\cos ny + t_{13}\cos nz = X_\sigma \\ t_{21}\cos nx + t_{22}\cos ny + t_{23}\cos nz = Y_\sigma \\ t_{31}\cos nx + t_{32}\cos ny + t_{33}\cos nz = Z_\sigma \end{array} \right\}, \qquad (3)$$

where n is the normal to the surface σ, the boundary of the elastic body, and X_σ, Y_σ, Z_σ are the components of the external forces acting on σ. These equations are not however sufficient to determine the general laws of elasticity; for this we must also make some hypotheses connecting the components γ_{rs} of the strains with the components t_{ik} of the stresses, and the simplest is to suppose that (at least to a first approximation) the γ_{rs}'s are *linear* functions of the t_{ik}'s. This hypothesis is called *Hooke's law*, because he first formulated it as early as 1660. We shall have therefore with this hypothesis

$$\gamma_{rs} = \sum_{ik} a_{rs,ik} t_{ik}; \qquad (4)$$

and solving with respect to the t_{ik}'s,

$$t_{ik} = \sum_{rs} b_{ik,rs} \gamma_{rs}; \qquad (5)$$

substituting these expressions in (2) and taking account of (1), we get a system of partial differential equations of the second order (of elliptic type) which, together with the boundary conditions (3), determines the components of the strains.

152. In the equations thus obtained we have however ignored the phenomenon of elastic fatigue, and, more generally, all *hereditary phenomena*. If these are to be taken into account, we must correct (4) into ([1])

$$\gamma_{rs} = \sum_{ik} a_{rs,ik} t_{ik} + \mathrm{F}_{rs}\left[\underset{-\infty}{\overset{t}{t_{11}}}, \underset{-\infty}{\overset{t}{t_{12}}}, \ldots ; t \right] \qquad (6)$$

by adding a corrective term F_{rs} which, as it depends on all the values of the stresses that have previously been acting on the body, will be a *functional* of the components $t_{ik}(\tau)$, which are themselves functions of the time τ considered in the interval from $-\infty$ to the instant in question t. We see therefore that

---

[1] Cf. VOLTERRA, (181).

it is only by means of the theory of functionals that it is possible to discuss exhaustively the phenomena of elasticity as they are in reality, i.e. depending on the previous history of the body. If in particular we suppose that the functionals F_{rs} can be expanded in series analogous to TAYLOR's series, and if we also extend to them the hypotheses about being *linear* functions (HOOKE's law), then the equations (6) will be reduced to the simpler form

$$\gamma_{rs} = \Sigma_{ik} a_{rs,ik} t_{ik} + \int_{-\infty}^{t} \Sigma \varphi_{rs,ik}(t,\tau) t_{ik}(\tau) d\tau; \quad (7)$$

and if the heredity before the moment 0 can be neglected, the integration can also be limited to the finite interval $(0, t)$. If the duration of the heredity is T_0 the integration can be limited to the interval $(t - T_0, t)$ (cf. No. 148). In the case of the closed cycle we shall have

$$\varphi_{rs,ik}(t,\tau) = \varphi_{rs,ik}(t - \tau).$$

Solving the system (7) of integral equations with respect to the t_{ik}'s and substituting as before in the equations (2), we shall get three partial derivative *integro-differential equations* of elliptic type to determine the three unknown functions u, v, w.

§ 4. *Hereditary Electromagnetism.*

153. Another similar application of functionals to hereditary phenomena is given by the theory of magnetism and electricity. With the classical hypotheses, in which heredity is not taken into account[1], the fundamental equations of electrodynamics are

$$a\frac{\partial P_m}{\partial t} = -\operatorname{rot} F_e, \quad a\frac{\partial P_e}{\partial t} = \operatorname{rot} F_m - 4\pi a C_e, \quad (1)$$

where P_m and F_m denote the vectors *magnetic induction* and *force*, and P_e, F_e, and C_e the vectors *electric induction, force, and current* respectively, and these vectors are connected by the linear relations

$$\begin{aligned}P_{e,x} &= \varepsilon_{11} F_{e,x} + \varepsilon_{12} F_{e,y} + \varepsilon_{13} F_{e,z} \\ P_{e,y} &= \varepsilon_{21} F_{e,x} + \varepsilon_{22} F_{e,y} + \varepsilon_{23} F_{e,z} \\ P_{e,z} &= \varepsilon_{31} F_{e,x} + \varepsilon_{32} F_{e,y} + \varepsilon_{33} F_{e,z}\end{aligned} \Bigg\}, \quad (2)$$

---

[1] Cf. HERTZ, (82).

VI APPLICATIONS

$$P_{m,x} = \mu_{11} F_{m,x} + \mu_{12} F_{m,y} + \mu_{13} F_{m,z}$$
$$P_{m,y} = \mu_{21} F_{m,x} + \mu_{22} F_{m,y} + \mu_{23} F_{m,z} \Big\}, \quad (3)$$
$$P_{m,z} = \mu_{31} F_{m,x} + \mu_{32} F_{m,y} + \mu_{33} F_{m,z}$$

$$C_{e,x} = \lambda_{11}(F_{e,x} - F'_{e,x}) + \lambda_{12}(F_{e,y} - F'_{e,y}) + \lambda_{13}(F_{e,z} - F'_{e,z})$$
$$C_{e,y} = \lambda_{21}(F_{e,x} - F'_{e,x}) + \lambda_{22}(F_{e,y} - F'_{e,y}) + \lambda_{23}(F_{e,z} - F'_{e,z}) \Big\}, (4)$$
$$C_{e,z} = \lambda_{31}(F_{e,x} - F'_{e,x}) + \lambda_{32}(F_{e,y} - F'_{e,y}) + \lambda_{33}(F_{e,z} - F'_{e,z})$$

where $F'_{e,x}$, $F'_{e,y}$, $F'_{e,z}$ are the components of the electro-motive force.

If however we wish to take heredity into account, and to discuss the phenomena of *magnetic hysteresis*, and others like them, we shall have to add to the linear expressions (2) and (3) a set of corrective terms analogous to those in the equations (6) of No. 152, which will be *functionals* of the functions $F_{e,x}(\tau)$, $F_{e,y}(\tau)$, $F_{e,z}(\tau)$, and $F_{m,x}(\tau)$, $F_{m,y}(\tau)$, $F_{m,z}(\tau)$ respectively, depending on the values taken by these functions when the time τ varies from $-\infty$ to the moment in question t([1]). In the case of *linear heredity* (which, however, in electrodynamics is insufficient to explain some phenomena)([2]), the terms to be added will be integrals analogous to those in (7) of No. 152. With these more precise hypotheses we naturally arrive at *integro-differential equations*, instead of at ordinary partial differential equations, as with the classic hypotheses.

154. Let us suppose the heredity linear, the medium homogeneous and isotropic, the case of the closed cycle, and $F'_e = 0$. Then the relations (2), (3), and (4) will be replaced by the relations

$$P_e(t) = \varepsilon F_e(t) + \int_0^{T_0} \Psi(\tau) F_e(t-\tau) d\tau$$
$$P_m(t) = \mu F_m(t) + \int_0^{T_0} \Phi(\tau) F_m(t-\tau) d\tau \Bigg\} .$$
$$C_e(t) = \lambda F_e$$

In these equations the coefficients of heredity $\Psi(\tau)$ and $\Phi(\tau)$ must be positive and decreasing, and must vanish for

---

([1]) Cf. VOLTERRA, (64) of Bibliography V.
([2]) Cf. VOLTERRA, (189), p. 249, Note 3; SERINI, (164); STRANEO, (166); KOSTITZIN, (93).

$\tau = T_0$, where the duration of the heredity is taken to be T_0, and where as primary histories we take those of electric force and magnetic force, and as hereditary histories those of the corresponding inductions.

Substituting the above expressions in the equations (1) we at once obtain the integro-differential equations which express the behaviour of electro-magnetic phenomena in the hereditary case[1].

§ 5. *Energy Equations.*

155. If hereditary effects are not present, equation (3) of No. 150 leads to the relation

$$d(T + E) = Q\dot{}dq = Qq' dt, \qquad (1)$$

where $T = \tfrac{1}{2} q'^2$ is the vis viva and $E = \tfrac{1}{2} bq^2$ the potential energy. We shall also have

$$E_m - E_m^0 = \int_0^t Q\, q'\, dt,$$

where we have put $E_m = T + E$ (the mechanical energy).

These are the fundamental energy equations of non-hereditary dynamics. In the hereditary case[2] we obtain instead of (1) the equation

$$d\left\{ \tfrac{1}{2} q'^2(t) + \tfrac{1}{2} mq^2(t) - \tfrac{1}{2}\int_0^{T_0} \Phi(\tau)\, [q(t) - q(t-\tau)]^2\, d\tau \right\}$$
$$+ \frac{dt}{2}\int_0^{T_0} \Phi'(\tau)\, [q(t) + q(t-\tau)]^2\, d\tau = Q(t)\, dq(t).$$

And since

$$\tfrac{1}{2} q'^2(t) = T, \quad \tfrac{1}{2} mq^2(t) - \tfrac{1}{2}\int_0^{T_0} \Phi(\tau)\, [q(t) - q(t-\tau)]^2\, d\tau = E_p$$

(cf. (4) of No. 150), and since we can put

$$E_\partial = \tfrac{1}{2}\int_0^{T_0} \Phi'(\tau)\, [q(t) - q(t-\tau)]^2\, d\tau,$$

we shall have

$$d(T + E_p) + E_\partial\, dt = Q(t)\, dq(t).$$

---

[1] Cf. GRAFFI, (72) and (73) of Bibliography V.
[2] Cf. VOLTERRA, (189).

APPLICATIONS

Now E_∂ is a positive quantity; hence the work of the external forces is always greater than the increment of the quantity

$$T + E_p.$$

If we write $T + E_p = E_m$ and integrate between two instants of time t_0 and t, it will follow from the preceding equation that

$$E_m - E_m^0 + \int_{t_0}^t E_\partial \, dt = L, \qquad (2)$$

where E_m and E_m^0 are the values of E_m at the times t and t_0, and L is the work of the external forces.

We shall by definition call E_p the *internal potential energy* and E_m the *mechanical energy*. We then have the principle of energy:

The work of the external forces always exceeds the variation of the mechanical energy by a positive quantity.

Now let us suppose that after a certain time the system returns to the initial conditions([1]) (i.e. initial from the hereditary point of view); then the mechanical energy will again take its initial value. Hence we have the result that *if at the end of a certain period of time the system returns to the same initial conditions from the hereditary point of view, the work done by the external forces is positive.* In this case nothing is changed from the mechanical point of view in the system, and hence from this point of view this positive work is dissipated work. It can at once be calculated from the formula (2) by taking $E_m = E_m^0$, which gives

$$\int_{t_0}^t E_\partial \, dt \qquad (3)$$

as the dissipated mechanical work. Naturally, according to the principles of the conservation of energy, *it must be transformed into other forms of energy.*

If the cycle described by the system is not closed, i.e. if the system does not return to the same conditions (from the

---

([1]) From the hereditary point of view the system is in the same conditions at two different instants when not only do the parameters that define the state of the system have the same values at the two instants, but also the values of these parameters are respectively equal to one another at the two intervals of time of amplitude T_0 which precede the two given instants.

hereditary point of view) at the time t as it was in at the time t_0, we cannot assert that the quantity (3) will always give us a quantity of mechanical work that is transformed into other forms of energy. It is in fact only *by definition* that E_p has been called the *internal potential energy* and E_m the *mechanical energy*. But we have demonstrated *the existence of a functional depending on the conditions of the system from the hereditary point of view and such that the work done by the external forces in the passage from one state to another is always greater than the variation of the functional.*

The functional considered is not the only one with these properties.

The different kinds of heredity and their various properties (cf. No. 148) lead to different forms of energy equations.

Thus if we consider the heredity subsequent to a certain instant, the energy relation previously obtained can be extended, and by reversing the heredity we can obtain a relation of a different form.

156. As an example of another form of energy relation we shall examine electromagnetic phenomena of hereditary character.

To determine the fundamental energy law we can apply Poynting's[1] well-known process to the equations (1) of No. 153. We must then calculate the sum of the scalar products

$$\frac{1}{4\pi}\frac{dP_e}{dt}\cdot F_e, \quad \frac{1}{4\pi}\frac{dP_m}{dt}\cdot F_m,$$

which, when all the necessary calculations and transformations have been made, is expressed by the formula

$$\frac{d}{dt}(E_e + E_m) + E_\partial,$$

where we have put

$$E_e = \frac{\varepsilon}{8\pi}F_e^2 + \int_0^{T_0}\frac{\Psi(\tau)}{8\pi}F_e^2(t-\tau)d\tau,$$

$$E_m = \frac{\mu}{8\pi}F_m^2 + \int_0^{T_0}\frac{\Phi(\tau)}{8\pi}F_m^2(t-\tau)d\tau,$$

$$E_\partial = -\frac{1}{8\pi}\int_0^{T_0}\{\Psi'(\tau)[F_e(t-\tau)-F_e(t)]^2 \\ + \Phi'(\tau)[F_m(t-\tau)-F_m(t)]^2\}d\tau,$$

---

[1] Cf. Poynting, (148).

VI APPLICATIONS

and where

$F_e^2(t)$, $F_m^2(t)$, $[F_e(t-\tau) - F_e(t)]^2$, $[F_m(t-\tau) - F_m(t)]^2$

denote respectively the squares of the tensors of the vectors

$F_e(t)$, $F_m(t)$, $F_e(t-\tau) - F_e(t)$, $F_m(t-\tau) - F_m(t)$.

If *by definition* we call E_e the *unit electric potential energy* and E_m the *unit magnetic potential energy*, while we call E_∂ the *unit electromagnetic energy of dissipation due to heredity*, these quantities will all be positive, and the flux of electromagnetic energy which crosses the boundary of a field S in an interval of time (t_0, t) will be equal to

$$\int_S (E_e + E_m) dS - \int_S (E_e^0 + E_m^0) dS + \int_{t_0}^t dt \int_S E_\partial dS + J,$$

where J denotes the Joule heat (measured in mechanical energy), and where the affix 0 has been used to denote the initial values (at the time t_0) of E_e and E_m. From this we get the following theorem: *the flux of the electromagnetic energy across the boundary of a field exceeds the sum of the Joule energy and the increment of the electromagnetic energy of the field by a positive quantity.*

The part dissipated (which in general is also transformed into heat) is calculated from the formula

$$\int_{t_0}^t dt \int_S E_\partial dS.$$

This assertion can be made at least whenever the system returns to its initial conditions from the hereditary point of view.

In this case also, therefore, we have demonstrated *the existence of a functional depending on the conditions of the system from the hereditary point of view, and such that its variations in any interval of time added to the Joule energy are always exceeded by the electromagnetic energy that enters the field from its boundary.*

This functional is not the only one with these properties. Another could in fact be calculated (analogous to the one found for dynamical systems) differing from the first in that only the values determining the state of the system during the period of duration of the heredity appear in the one, while the other contains the differences between the elements defining the state of the system at the present moment and those determining it in the past.

In the case of dynamics the energy principles can also be extended even if the heredity is not linear.

Section V.
§ 1. *Functional Operators.*

157. A point of view fundamentally different from the one we have been following in questions of functional calculus is that of the so-called *symbolic calculus*, where the tendency is essentially algorismic and formal. Among its many devotees, we may mention LEIBNIZ, LAGRANGE, SERVOIS, and many others. The history of these researches has been given by TARDY[1].

This field of research (*functional operations*) also includes the study of derivatives with fractional indexes, referred to above in Chapter II (No. 49).

158. O. HEAVISIDE continuously and systematically uses functional operators in his important researches in electrodynamics[2]. G. GIORGI has rigorously applied the methods of functional operators in a great number of papers, obtaining interesting results relating to variable electric currents, and generally for the integration of differential equations[3].

The modern theories of physics make much use of functional operators[4].

159. But the principal development of this calculus, so far as concerns linear functional operations (or "distributive operations") in the field of analytic functions, is due to the work of PINCHERLE[5] in his numerous writings which it is impossible to recapitulate here. It is already several years since CALÒ[6] demonstrated the relations[7] between quadratures and PINCHERLE's distributive operations, which, as we have said, are linear functionals defined in the field of analytic functions; they retain the property of being analytic with respect to the

---

[1] Cf. TARDY, (220).
[2] Cf. HEAVISIDE, (209), (210).
[3] Cf. GIORGI, (74)—(77), (202), (203).
[4] Cf. BIRTWISTLE, (198).
[5] Cf. PINCHERLE, (140)—(144), (217).
[6] Cf. CALÒ, (18).
[7] Cf. also FANTAPPIÈ, (34).

parameters occuring in the functions to which these functionals are applied.

As M. WINTER has observed, these ideas relative to the formal development of the properties of operators, which are related to PINCHERLE's researches, are of a wholly different nature to the other current of conceptions, consisting in the generalisation of the fundamental notions of the differential and integral calculus, which forms the subject of the present book([1]).

160. The interesting researches of BOURLET([2]) have reference to functional operations, called by him *functional transmutations*. He applies a functional transmutation to any two functions whatever and to their sum, and considers the problem of obtaining the transmutation from the relations between them.

§ 2. *Analytic Functionals.*

161. L. FANTAPPIÈ([3]) has recently reduced the study of linear analytic functionals $F[y(t)]$([4]) to the study of a function $v(\alpha)$ of one complex variable (the *functional indicatrix* of F), which, when the functional F is given, is completely determined by

$$v(\alpha) = F\left[\frac{1}{t-\alpha}\right], \tag{1}$$

and which in turn determines the functional by means of the formula

$$F[y(t)] = \frac{1}{2\pi\iota}\int_C v(t)\,y(t)\,dt, \tag{2}$$

where C is any closed curve whatever on the complex sphere *which contains in its interior all the singular points of $y(t)$ but leaves outside all those of $v(t)$*. FANTAPPIÈ gives the name *hemi-symmetrical functional product* to the second member of (2). If F depends on a parameter z, i.e. $F[y(t); z] = f(z)$, v also depends on z, i.e. $v = v(z, \alpha)$. It is easy to see that the singular points

---

([1]) Cf. WINTER, (92) of Bibliography I.
([2]) Cf. BOURLET, (16).
([3]) Cf. FANTAPPIÈ, (35).
([4]) F is an analytic functional if $F[y(t, \alpha)]$ is an analytic function of the parameter α, when $y(t, \alpha)$ is an analytic function of t and of the parameter α.

of $f(z)$ are those of the functions $v(z, t_1), v(z, t_2), \ldots,$ if $t_1, t_2, \ldots$ are the singular points of $y(t)$. As a particular case of this theorem we have HADAMARD's well-known theorem, already mentioned in No. 103, which gives the positions of the singular points of PINCHERLE's "normal operations", and another analogous theorem which gives the positions of the singular points of the functions that are obtained by applying distributive operations that are permutable with derivation, etc.

The variation $\delta \, \mathrm{F}[y(t)]$ of a non-linear functional $\mathrm{F}[y(t)]$ corresponding to a variation δy of the variable function is a linear functional of δy. FANTAPPIÈ takes the functional indicatrix of this linear functional as a definition of the functional derivative of F.

Having found the rules of the calculus of functional derivatives, FANTAPPIÈ has succeeded in expressing an analytic functional, in the domain of a given function, by means of a series analogous to TAYLOR's series. The expansions of VOLTERRA and FRÉCHET (cf. No. 25) are particular cases of this series.

From this general expression of an analytic functional we can pass to the problem of solving a functional equation

$$\mathrm{F}[y(t), z] = f(z)$$

in the analytical domain.

FANTAPPIÈ gives the solution $y(t)$ in the form of a series of the preceding type, calculating its successive terms by solving a certain linear differential equation whose kernel is a function of two variables([1]).

This provides an instance of the usefulness of the study of analytic functionals of two variables.

If

$$v(\alpha, \beta) = \mathrm{F}\left[\frac{1}{(t_1 - \alpha)(t_2 - \beta)}\right]$$

is the functional indicatrix of the linear functional $F[y(t_1, t_2)]$, we can calculate this functional by means of a double integral depending on v; the integration can however be taken over several surfaces and we can thus obtain different values for the integral. It follows that a linear functional of a function y of two variables can be *many-valued* even when the functional in-

---

([1]) Cf. FANTAPPIÈ, (46), (48).

dicatrix v and the function y are *single-valued*. This establishes a fundamental difference between analytic functionals of a function of one variable and analytic functionals of a function of two variables.

If the singular curves of y are tangential to the singular curves of the indicatrix, y can be a *branch-function*. Hence FAN- TAPPIÈ obtains a theorem for double power-series analogous to HADAMARD's theorem (No. 103)([1]).

162. By applying the theory of analytic functionals FANTAPPIÈ has modified some methods employed in *Quantum Mechanics*, in which series of the types $\sum_{k}^{\infty} a_{rk} b_{ks}$ (the product of two matrices), $\sum_{n}^{\infty} a_{nn}$ (the diagonal sum of a matrix), and $\sum_n f_n\, a_{rs}^{(n)}$ (a function $f(A)$ of a matrix) are considered. These series have sometimes been used without verifying their convergence, and using, by analogy, HILBERT's theory of forms of an infinite number of variables even when it is not applicable. FANTAPPIÈ substitutes for HILBERT's expression $\sum a_n\, x_n$ the expression

$$\frac{1}{2\pi\iota}\int_c a(t)\,\frac{x\left(\dfrac{1}{t}\right)}{t}\,dt$$

(a *symmetrical functional product*), where

$$a(t) = \sum a_n t^n,\quad x(t) = \sum x_n t^n;$$

this is equal to $\sum a_n x_n$ when this series converges, but has a meaning even when it does not converge, it being supposed that the singular points of $a(t)$ and $x(t)$ are not reciprocals. It is therefore not necessary to verify that the series $\sum a_n^2$, $\sum x_n^2$ are convergent, but only that $\sqrt[n]{|a_n|}$ and $\sqrt[n]{|x_n|}$ are always less than some finite number.

For the infinite matrix $\|a_{rs}\|$ which appears in the Quantum Theory FANTAPPIÈ also substitutes the corresponding function of two variables $a(z_1, z_2) = \sum_{rs} a_{rs}\, z_1^r z_2^s$.

For the product of two matrices $\|a_{rs}\|\,\|b_{rs}\|$ he substitutes the symmetrical product by composition of the corresponding functions, and for the matrix function of another matrix he sub-

---

([1]) Cf. FANTAPPIÈ, (36)—(45), (51), (52). These papers will be included in a memoir to be published in the *Memorie della R. Accademia dei Lincei*, Series VI, Vol. 3.

stitutes what he calls the *series by symmetrical functional composition*. He thus avoids the use of divergent series. It is to be noted that the calculation of the diagonal sum leads him to many-valued expressions([1]).

§ 3. *Abstract Spaces.*

164. FRÉCHET is the author of the idea of considering, not the functional derivative, but the *differential* of a functional, and then passing to the functional derivative([2]); in VOLTERRA's early works([3]), on the contrary, the method was to start from the functional derivative in order to arrive at the differential, and this method has also been adopted in these lectures. At bottom, FRÉCHET maintains the utility in the theory of functionals of *taking the function, from the very outset and directly, as the independent variable,* rather than of considering the functional as the limit of a function of an infinite number of variables. It must however be noted that in practice it was the primitive concept that led to VOLTERRA's first results, and this also happened later on in the case of FREDHOLM and HILBERT, who have since followed the same road([4]). Now, however, it is useful to follow the new road as well, which leads to the more recent concept of a functional due to FRÉCHET and the writers belonging to the same school.

164. FRÉCHET's new idea is that *it is not necessary to take into account the nature of the variable,* whether it is a number, a line, a function, or anything else; and he has in fact proposed to extend the infinitesimal calculus to the case in which the variable is of any nature whatever. In this he has been gradually approaching the ideas that MOORE([5]) has been developing in America since 1908, which belong to what is called "general analysis". "General analysis" proposes, in fact, to extract more abstract common notions from known theories, and to generalise these theories by removing from them any particular properties that are related to the concrete elements on which they are based. FRÉCHET gives an example of this in his vector theory; and it

---
([1]) Cf. FANTAPPIÈ, (47), (50).
([2]) Cf. FRÉCHET, (31) of Bibliography I.
([3]) Cf. VOLTERRA, (83) of Bibliography I.
([4]) Cf. FREDHOLM, (20), and HILBERT, (23), of Bibliography II.
([5]) Cf. MOORE, (128).

is what mathematics has always done, passing continually from the concrete to the abstract.

The new concept of functionals, therefore, deprives the variable of all concrete character. If however, we wish to create an infinitesimal calculus, we must also give very small changes to this abstract variable; hence the necessity of transporting the concept of distance into these new fields of abstract variables (cf. Nos. 13 to 19). It is therefore necessary in the first place to examine these fields, and study the infinitesimal properties of their aggregates of points.

General analysis and functional analysis, however, need from their very earliest stages those notions which in the ordinary theory of functions have been utilised only for its latest developments, as, for example, that of semicontinuity, which was first applied to the case, mentioned above (No. 132), of the area of a surface as the limit of a polyhedron.

For abstract aggregates, the first thing to do is therefore, as we have said, to introduce the notion of "distance", which we have already given in Chapter I, § 4, or the more general notion of "interval", in order to be able to define the "derived" aggregate of a given one. More generally, however, it is possible to consider abstract aggregates in which the operation of derivation of the aggregates is defined in any way whatever, without having recourse to other intermediary notions (distance, interval, FRÉCHET's "neighbourhood", etc.). The term "topological space" is applied to any collection of elements of the same kind for which a definition is given of the derived aggregates, and given in such a way that the derived aggregates are also contained in the same space[1].

Very particular cases of topological spaces are the "vectorial spaces". These are spaces in which to every pair of elements or points there can be associated an element of a different kind having the essential properties of vectors.

Among the latest work on this subject should be mentioned that of URYSOHN[2], of ALEXANDROFF[3], a note by LÉVY[4] on probabilities in abstract aggregates, and many others[5].

[1] Cf. FRÉCHET, (57)—(60), (204)—(208); SIERSPINSKI, (165).
[2] Cf. URYSOHN, (172).
[3] Cf. ALEXANDROFF, (2).
[4] Cf. LÉVY, (110).
[5] FRÉCHET's book, *Les espaces abstraits* (208), contains a bibliography of the subject.

§4. Various Applications.

165. Recent applications of the theory of functionals have been made in the most varied fields; we may mention its applications to ballistics by BLISS[1], and to political economy[2] and statistics by RISSER[3], SCHÖNBAUM[4], EVANS[5], HOTELLING[6], and ROOS[7].

166. An interesting application of integro-differential equations is to be found in the biologico-mathematical theory of the fluctuations of species living together[8].

If two species have coefficients of increase μ_1 and μ_2, and if the numbers of individuals of the species at time t are N_1 and N_2 respectively, we have

$$\frac{dN_1}{dt} = \mu_1 N_1, \quad \frac{dN_2}{dt} = \mu_2 N_2. \tag{1}$$

If $\mu_1 = \varepsilon_1$, a positive constant, $\mu_2 = \varepsilon_2$, a negative constant, the first species increases and the second diminishes exponentially; but if the second species can feed on the individuals of the first, μ_1 and μ_2 will not be constants, because μ_1 will diminish as N_2 increases, and μ_2 will increase as N_1 increases. If therefore we take

$$\mu_1 = \varepsilon - \gamma_1 N_2, \quad \mu_2 = -\varepsilon_2 + \gamma_2 N_1,$$

where γ_1 and γ_2 are positive constants, and substitute these values in (1), we have two differential equations which lead to the general laws of the fluctuations of two species living together, one of which feeds on the other.

Now the coefficient of increase μ_2 at any moment does not depend only on the amount of food found at that moment by the second species (which is a function of the value at the moment of the number N_1), but also on the food previously found, which depends, by a functional law, on all the values previously taken by N_1.

---

[1] Cf. BLISS, (11).
[2] L. AMOROSO has made some remarks on this subject; cf. (3).
[3] Cf. RISSER, (151).
[4] Cf. SCHÖNBAUM, (161), (162).
[5] Cf. EVANS, (28), (29).
[6] Cf. HOTELLING, (88).
[7] Cf. ROOS, (152)—(157).
[8] Cf. VOLTERRA, (187).

APPLICATIONS

If we wish to take this fact into account we find that we have to deal with functional questions of the kind we have called *hereditary* (cf. Chapter V, No. 115), and the second of the equations (1) must be replaced by an integro-differential equation.

In order to treat the question analytically in a still more general and symmetrical way, we can replace the two equations (1) by two integro-differential equations, which, on the hypothesis of heredity of duration T_0, take the form

$$\frac{dN_1}{dt} = N_1(t) \left\{ \varepsilon_1 - \gamma_1 N_2(t) - \int_{t-T_0}^{t} F_1(t-\tau) N_2(\tau) \, d\tau \right\},$$

$$\frac{dN_2}{dt} = N_2(t) \left\{ -\varepsilon_2 + \gamma_2 N_1(t) + \int_{t-T_0}^{t} F_2(t-\tau) N_1(\tau) \, d\tau \right\}.$$

It can be shown that the integrals of these equations continue to exist for infinitely large values of the time; that a stationary state exists, and that the numbers of individuals of the two species oscillate indefinitely about the numbers corresponding to the stationary state, which are their asymptotic means. Lastly it is possible to find the laws of perturbation of these means, and it can be shown that small periodical fluctuations are not possible.

BIBLIOGRAPHY VI.([1])

(1) ADAMS (C. R.).—Note on integro-q-difference equations. (*Trans. of American Math. Soc.*, Vol. 31, 1929.)
(2) ALEXANDROFF (P.).—Sur les ensembles complémentaires aux ensembles A. (*Fundamenta Mathematica* (Warsaw), Vol. V, 1924.) For other papers by this author see (208).
(3) AMOROSO (L.).—Le equazioni differenziali della dinamica economica. (*Atti. Congr. Int. Mat. Bologna*, 1928.)
(4) ANDREOLI (G.).—Sulla calibrazione dei tubi termometrici e le equazioni funzionali. (*Atti R. Ist. Veneto*, Vol. 74, 1914-15.)
(5) BANACH (P.).—Sur les opérations dans les ensembles abstraits et leur application aux équations intégrales. (*Fundamenta Mathematica*, Vol. III, 1922.)
(6) BÄR (R.).—Über Greensche Randwertaufgaben bei der Schwingungsgleichung. (*Math. Annalen*, Vol. LXXVIII, 1917-18.)
(7) BARNETT (J. A.).—Functionals invariant under one-parameter continuous groups of transformations in the space of continuous functions. (*Proc. N. Ac. Sc. (U. S. A.)*, Vol. 6, 1920.)
(8) BATEMAN (H.).—On integral equations occurring in a mathematical theory of retail trade. (*Messenger of Math.*, Vol. 49.)
(9) BERTRAND (G.).—La théorie des marées et les équations intégrales. (*Ann. Ec. Normale de Paris*, Vol. 40, 1923.)
(10) BIRKHOFF-KELLOGG.—Invariant points in function space. (*Trans. of American Math. Soc.*, Vol. XXIII, 1922.)
(11) BLISS (G. A.).—Differential equations containing arbitrary functions. Functions of lines in ballistics. (*Trans. of American Math. Soc.*, Vol. XXI, 1920.)
(12) BOLTZMANN (L.).—Zur Theorie der elastischen Nachwirkung. (*Wien. Ber.*, 70; *Wiss. Abh.*, Vol. I, 1874.)
(13) BONFERRONI.—Teoria degli accumuli e legge di capitalizzazione. (*Giornale di Mat. Finanziaria*, Vol. VI.)
(14) BORN (M.) and HEISENBERG (W.).—La mécanique des quanta. (*Rapp. Disc. 5e Cons. Physique*, Inst. Solvay, Brussels, 1928.)
(15) BOULIGAND (G.).—Fonctions harmoniques. Principes de Picard et de Dirichlet. (*Memorial des Sci. Math.*, Fasc. XI, 1926.)
(16) BOURLET (C.).—Sur les opérations en général et les équations différentielles linéaires d'ordre infini. (*Ann. Ec. Normale de Paris*, 3rd series, Vol. XIV, 1897.)
(17) BRAY (H. E.) and EVANS (G. C.).—A class of functions harmonic within the sphere. (*Amer. Journ. Math.*, Vol. 49, 1927.)

([1]) Bibliographies V and VI contain only a concise bibliography of integro-differential equations.

(18) Calò (B.).—Sulle operazioni funzionali distributive. (*R. Acc. dei Lincei. Rend.*, Vol. IV, Series 5, 1895.)
(19) Chrystal (G.).—On the hydrodynamical theory of seiches. (*Trans. of R. Soc. Edinburgh*, Vol. XLI, 1905.)
(20) Courant (R.).—Über die Abhängigkeit der Schwingungszahlen einer Membran von ihrer Begrenzung und über asymptotische Eigenwertverteilung. (*Gött. Nachr.*, 1919.)
(21) Davis (H. T.).—Relating to the proof of an existence theorem for a certain type of boundary value problem. (*American Math. Soc. Bull.*, 2nd series, Vol. 28, 1922.)
(22) Delsarte (J.).—Sur les transformations fonctionnelles et les rotations fonctionnelles non-euclidiennes. (*C. R. Acad. Sci. Paris*, T. 186, 1928.)
(23) Delsarte (J.).—Les rotations fonctionnelles. (*Ann. Fac. Sciences Toulouse*, 1928.)
(24) Delsarte (J.).—Mémoire sur les groupes finis de rotations fonctionnelles. (*Rend. Circ. Mat. Palermo*, 1929.)
(25) Dines (L. L.).—Projective transformations in function space. (*American Math. Soc. Trans.*, Vol. 20, 1919.)
(26) Evans (G. C.).—Fundamental points of potential theory. (*Rice Inst. Pamphlet*, Vol. 7, 1920.)
(27) Evans (G. C.).—The dynamics of monopoly. (*Am. Math. Monthly*, Vol. XXXI, 1924.)
(28) Evans (G. C.).—Economics and calculus of variations. (*Proc. Nat. Acad. Sci.*, Vol. 2, 1925.)
(29) Evans (G. C.).—The mathematical theory of economics. (*Amer. Math. Monthly*, Vol. 32, 1925.)
(30) Evans (G. C.).—Generalized Neumann problems for the sphere. (*Amer. Journ. Math.*, Vol. 50, 1928.)
(31) Evans (G. C.).—Note on a theorem of Bôcher. (*Amer. Journ. Math.*, Vol. 50, 1928.)
(32) Evans (G. C.).—Discontinuous boundary value problems of the first kind for Poisson's equation. (*Amer. Journ. Math.*, Vol. 51, 1929.)
(33) Evans (G. C.). and Miles (E. R. C.).—Potentials of general masses in single and double layers. (*Proc. Nat. Acad. Sci.*, Vol. 15, 1929.)
(34) Fantappiè (L.).—Sulla riduzione delle operazioni distributive di Pincherle alle funzionali lineari di Volterra. (*R. Acc. dei Lincei. Rend.*, Vol. I, Series 6, 1925.)
(35) Fantappiè (L.).—Le funzionali lineari analitiche e le loro singolarità. (*R. Acc. dei Lincei. Rend.*, Vol. I, Series 6, 1925.)
(36) Fantappiè (L.).—La derivazione delle funzionali analitiche. (*R. Acc. dei Lincei. Rend.*, Vol. I, Series 6, 1925.)
(37) Fantappiè (L.).—Le operazioni distributive esprimibili con un numero finito di operazioni elementari. (*Boll. Un. Mat. Ital.*, Anno IV, No. 4, 1925.)
(38) Fantappiè (L.).—Risoluzione di una classe di equazioni integrali di la specie a limiti costanti. (*R. Acc. dei Lincei. Rend.*, Vol. 2, Series 6, 1925.)

(39) FANTAPPIÈ (L.).—Determinazione dei gruppi a un parametro di funzionali lineari (*R. Acc. dei Lincei. Rend.*, Vol. III, Series 6, 1926.)
(40) FANTAPPIÈ (L.).—I funzionali analitici non lineari. (*R. Acc. dei Lincei. Rend.*, Vol. III, Series 6, 1926.)
(41) FANTAPPIÈ (L.).—La polidromia dei funzionali analitici lineari. (*R. Acc. dei Lincei. Rend.*, Vol. IV, Series 6, 1926.)
(42) FANTAPPIÈ (L.).—Les fonctionelles analytiques qui sont des fonctions d'un nombre fini de fonctionnelles linéaires. (*C. R. de l'Acad. des Sciences de Paris*, Vol. 183, 1926.)
(43) FANTAPPIÈ (L.).—Sur une classe de fontionnelles analytiques. (*C. R. de l'Acad. des Sciences de Paris*, Vol 183, 1926.)
(44) FANTAPPIÈ (L.).—I funzionali analitici. (*Rend. del Sem. Mat. della R. Univer. di Roma*, Series 2, Vol. IV, 1926.)
(45) FANTAPPIÈ (L.).—La teoria dei funzionali analitici nell' integrazione delle equazioni lineari a derivate parziali di qualsiasi ordine. (*R. Acc. dei Lincei. Rend.*, Vol. IV, Series 6, 1926.)
(46) FANTAPPIÈ (L.).—I funzionali analitici delle funzioni di due variabili complesse. (*R. Acc. dei Lincei. Rend.*, Vol. V, Series 6, 1927.)
(47) FANTAPPIÈ (L.).—Le calcul des matrices. (*C. R. de l'Acad. des Sciences de Paris*, Vol. 186, 1928.)
(48) FANTAPPIÈ (L.).—I funzionali lineari delle funzioni di due variabili complesse. (3 notes.) (*R. Acc. dei Lincei. Rend.*, Vol. VII, Series 6, 1928.)
(49) FANTAPPIÈ (L.).—Sobre un nuevo determinante funcional. (*Rev. Mat. Hispano-Amer.*, 1928.)
(50) FANTAPPIÈ (L.).—Gli operatori funzionali e il calcolo delle matrici infinite nella teoria dei quanti. (2 notes.) (*R. Acc. dei Lincei. Rend.*, Vol. VIII, Series 6, 1928; Vol. IX, Series 6, 1929.)
(51) FANTAPPIÈ (L.).— Cenni riassuntivi sulla teoria dei funzionali analitici. (*Studia Math.*, Vol. 1, 1929.)
(52) FANTAPPIÈ (L.).—Le equazioni funzionali lineari nel campo complesso. (*R. Acc. dei Lincei. Rend.*, Vol. IX, Series 6, 1929.)
(53) FOREL (M.).—Le Léman. Monographie limnologique. Vol. II. Lausanne, 1892-1895.
(54) FRANK (PH.).—Die Integralgleichungen in der Theorie der kleinen Schwingungen von Fäden und das Rayleigh'sche Prinzip. (*Sitz. Ak. der Wissensch. Wien*, Vol. 117, 1908.)
(55) FRANK (PH.) and PICK (G.).—Sur quelques mesures dans l'espace fonctionnel. (*C. R. de l'Acad. des Sciences de Paris*, Vol. 158, 1914.)
(56) FRANK (PH.) and PICK (G.).—Distanzschätzungen im Funktionenraum. (*Math. Ann.*, 1915.)
(57) FRÉCHET (M.)—Essai de géométrie analytique à une infinité de coordonnées. (*Nouv Ann. de Math.*, Vol. VIII, 1908.)
(58) FRÉCHET (M).—Les ensembles abstraits et le calcul fonctionnel. (*Rend. del Cir. Mat. di Palermo*, Vol. XXX, 1910.)
(59) FRÉCHET (M.).—Sur la notion de voisinage dans les ensembles abstraits (*Bull. Soc. Math.*, Series 2, Vol. XLII, 1918.)
(60) FRÉCHET (M.)—Sur l'homéomorphie des ensembles dénombrables. (*Bull Ac. polonaise*, 1920.)

VI BIBLIOGRAPHY 211

(61) FRÉCHET (M.).— Sur les ensembles abstraits. *(Ann. Ec. Normal de Paris*, 1922.)
(62) FRÉCHET (M.).— Esquisse d'une théorie des ensembles abstraits. (*Sir Asutosh Mooregee's Comm.*, Vol. II; The Baptist Mission Press, Calcutta, 1922.)
(63) FRÉCHET (M.).— Des familles et fonctions additives d'ensembles abstraits. *Fundamenta Math.*, Vol. IV, 1923; Vol. V, 1924.)
(64) FRÉCHET (M.).— Sur la distance de deux ensembles. *(Bull. of Calcutta Math. Soc.*, Vol XV, 1923-24).
(65) FRÉCHET (M.).—I. Sur l'aire des surfaces polyédrales.—II. Sur la distance de deux surfaces. (*Ann. Soc. Polonaise de Math.* (Cracow), 1924.)
(66) FRÉCHET (M.).— Sur la terminologie de la théorie des ensembles abstraits. *(C. R. Congrès des Soc. Sav.*, 1924.)
(67) FRÉCHET (M.).—L'analyse générale et les ensembles abstraits. (*Rev. de Métaph. et Mor.*, Vol. 32, 1925.)
(68) FRÉCHET (M.).—La notion de différentielle dans l'analyse générale. (*Ann. Ec. Normale de Paris*, Series 3, Vol. 41, 1925.)
(69) FRÉCHET (M.). — Démonstration de quelques propriétés des ensembles abstraits. (*Am. Journ. Math.*, Vol. L, 1928.)
(70) FRÉCHET (M.).—Quelques propriétés des ensembles abstraits. (*Fund. Math.*, Vols. X, XII, 1927, 1928.)
(71) FREDHOLM (I.).—Sur une nouvelle méthode pour la résolution du problème de Dirichlet. (*K. Vet. Akad.* (Stockholm), 1900.)
(72) FREDHOLM (I.).—Sur la réduction d'un problème de la mécanique rationelle à une équation intégrale linéaire. (*C. R. de l'Acad. des Sciences de Paris*, Vol. 171, 1920.)
(73) FREDHOLM (I.).—Une application de la théorie des équations intégrales. (*C. R. de l'Acad. des Sciences de Paris*, Vol. 174, 1922.)
(74) GIORGI (G.).—Il metodo simbolico nello studio delle correnti variabili. (*Atti Assoc. Elettrotecnica Italiana*, 11 Oct. 1903.)
(75) GIORGI (G.).—Sul calcolo delle soluzioni funzionali originate dai problemi di elettrodinamica. (*Atti Assoc. Elettrotecnica Italiana*, Vol. IX, 1905.)
(76) GIORGI G.).—Sui problemi della elasticità ereditaria. (*R. Acc. dei Lincei. Rend.*, Vol. XXI, Series 5, 1912.)
(77) GIORGI (G.).—Sugli operatori funzionali ereditari. (*R. Acc. dei Lincei. Rend.*, Vol. XXI, Series 5, 1912.)
(78) HADAMARD (J.).—Les problèmes aux limites dans la théorie des équations aux dérivées partielles. (*Bull. Soc. Franç. de Physique*, 1906.)
(79) HADAMARD (J.).—Le développement et le rôle scientifique du calcul fonctionnel. (*Atti Congr. Int. Mat.*, Bologna, 1928, Vol. I.)
(80) HECKE (E.).—Über die Integralgleichung der kinetischen Gastheorie. (*Math. Zeitschr.*, Vol. 12, 1922.)
(81) HEISENBERG (W.) and PAULI (W.).—Zur Quantendynamik der Wellenfelder. (*Zeitschr. f. Physik*, Vol. 56, 1929.)
(82) HERTZ (H.).—Über die Grundgleichungen der Elektrodynamik für ruhende Körper. (*Wiedemann's Annalen*, 40, p. 577; *Gesammelte Werke*, Vol. II, p. 208.)

212 BIBLIOGRAPHY CHAP.

(83) HILBERT (D.).—Über das Dirichlet'sche Prinzip. (*Jahresber. Deutsch. Math. Verein*, 1900.)
(84) HILBERT (D.).—Über das Dirichlet'sche Prinzip. (*Festschr. Feier 150-jähr. Best. K. Ges. Wiss. Göttingen*, 1901.)
(85) HILBERT (D.).—Wesen und Ziele einer Analysis der unendlich vielen Variabeln. (*Rend. del Cir. Mat. di Palermo*, Vol. XXVII, 1908.)
(86) HIRAHAWA (H.).—On a simple integral equation. (*Tôhoku Math. Journ.*, Vol. 8, 1915-16.)
(87) HOSTINSKY (B.).—Les fonctions fondamentales du problème de Dirichlet. (*Publ. de la Fac. des Sciences de l'Université Masaryk*, Brno, 1924.)
(88) HOTELLING (H.).—A general mathematical theory of depreciation (*Journ. Amer. Statistical Assoc.*, 1925.)
(89) KELLOGG.—Invariant points in function space [cf. BIRKHOFF, (10)].
(90) KORN (C.).—Über die beiden bisher zur Lösung der ersten Randwertaufgabe der Elastizitätstheorie eingeschlagenen Wege. (*Annaes da Ac. Polyt. do Porto*, Ext. do Vol. X, Coimbra, 1915.)
(91) KOSTITZIN (V. A.).—Sur les solutions singulières des équations intégrales du cycle fermé. (*Rec. Math. Moscou*, Vol. 33, 1926.)
(92) KOSTITZIN (V. A.).—Sur les solutions singulières des équations intégrales de Volterra. (*C. R. Acad. Sci. Paris*, Vol. 184, 1927.)
(93) KOSTITZIN (V. A.).—Sur quelques applications des équations intégrales au problème de l'hystérésis magnétique. (*Proc. Int. Math. Congr.*, Toronto, Vol. II, 1928.)
(94) KOWALEWSKI (G.).—Les formules de Frenet dans l'espace fonctionnel. (*C. R. de l'Acad. des Sciences de Paris*, Vol. 151, 1910.)
(95) KOWALEWSKI (G.).—Über Funktionenräume. I.-II. Mitteil. (*Sitz. K. Ak. in Wien*, Vol. 120, 1911.)
(96) KOWALEWSKI (G.).—Sur une propriété des transformations de Volterra. (*C. R. de l'Acad. des Sciences de Paris*, Vol. 153, 1911.)
(97) KOWALEWSKI (G.).—Sur une classe de transformations infinitésimales de l'espace fonctionnel. (*C. R. de l'Acad. des Sciences de Paris*, Vol. 153, 1911.)
(98) LAURICELLA (G.).—Sulla risoluzione del problema di Dirichlet col metodo di Fredholm. (*R. Acc. dei Lincei. Rend.*, Vol. XV, Series 5, 1906.)
(99) LAURICELLA (G.).—Applicazione della teoria di Fredholm al problema del raffreddamento dei corpi. (*Ann. di Mat.*, Vol. XIV, Series 3, 1907.)
(100) LAURICELLA (G.).—Alcune applicazioni della teoria delle equazioni funzionali alla fisica matematica. (*Nuovo Cim.*, Series 5, Vol. XIII, 1907.)
(101) LAURICELLA (G.).—Sull' equazione integrale di 1a specie relativa al problema di Dirichlet sul piano. (*R. Acc. dei Lincei. Rend.*, Vol. XIX, Series 5, 1910.)
(102) LAURICELLA (G.).—Sulla risoluzione delle equazioni integro-differenziali dell' equilibrio dei corpi elastici isotropi per dati spostamenti in superficie. (*R. Acc. dei Lincei. Rend.*, Vol. XXI, Series 5, 1912.)
(103) LENSE (T.).—Über eine Integralgleichung in der Theorie der heterogenen Gleichgewichtsfiguren. (*Math. Zeitschr.*, Vol. 16, 1923.)

(104) LEVI (E. E.).—Sur l'application des équations intégrales au problème de Riemann. (*Nachr. K. Ges. zu Göttingen*, 1908.)
(105) LEVI-CIVITA (T.).—Sui gruppi di operazioni funzionali. (*R. Ist. Lombardo*, Series 2, Vol. XXVIII, 1895.)
(106) LEVI-CIVITA (T.).—I gruppi di operazioni funzionali e l'inversione degli integrali definiti. (*R. Ist. Lombardo*, Series 2, Vol. XXVIII, 1895.)
(107) LEVI-CIVITA (T.).—Alcune osservazioni alla nota: Sui gruppi di operazioni funzionali. (*R. Ist. Lombardo*, Series 2, Vol. XXVIII, 1895.)
(108) LÉVY (P.).—Sur une généralisation de la méthode de Fredholm pour la résolution du problème de Dirichlet. (*Journal de l'École Polyt.*, 1911.)
(109) LÉVY (P.).—Sur la variation de la distribution de l'electricité sur un conducteur dont la surface se déforme. (*Bull. Soc. Math. de France*, Vol. 46, 1918.)
(110) LÉVY (P.).—Calcul des probabilités. Note: Les lois de probabilité dans les ensembles abstraits. (Paris, Gauthier-Villars, 1925.)
(111) LICHTENSTEIN (L.).—Untersuchungen über die Gestalt der Himmelskörper. (*Math. Zeitschr.*, Vols. 10, 12, 13, 17, 1921—1923.)
(112) LICHTENSTEIN (L.).—Über die erste Randwertaufgabe der Elastizitätstheorie. (*Math. Zeitschr.*, Vol. 20, 1924.)
(113) LICHTENSTEIN (L.).—Über ein. existenzprobl. d. Hydr. homog. unzusammendrückb. reibungsloser Fluss. und d. Helmholtzsch. Wirbelsätze. (*Math. Zeitschr.*, Vol. 23, 1925.)
(114) MANNEBACK (CH.).—An integral equation for skin effect in parallel conductors. (*Cambridge (Mass.) Journ. Math. Phys.*, Vol. I, 1922.)
(115) MARCOLONGO (R.).—La teoria delle equazioni integrali e le sue applicazioni alla fisica matematica. (*R. Acc. dei Lincei. Rend.*, Vol. XVI, Series 5, 1907.)
(116) MARCOLONGO (R.).—La théorie des équations intégrales et ses applications à la physique mathématique. (*Ann. Fac. des Sciences de Toulouse*, Series 2, Vol. IX.) (Translation of the preceding paper.)
(117) MARIA (A. J.).—Functions of plurisegments. (*Trans. of American Math. Soc.*, Vol. 28, 1926.)
(118) MATTEUZZI (L.).—Sulla determinazione delle seiches forzate e delle seiches libere mediante una equazione integrale di Volterra di 2a specie. (*R. Acc. dei Lincei. Rend.*, Vol. XXXIII, Series 5, 1924.)
(119) MÉRIAN (I. R.).—Über die Bewegung tropfbarer Flüssigkeiten in Gefäßen. (*Math. Ann.*, Vol. XXVIII, 1885.)
(120) MICHAL (A. B.).—Integro-differential invariants of one-parameter groups of Fredholm transformations. (*Bull. American Math. Soc.*, Vol. 30, 1924.)
(121) MICHAL (A. B.).—Functionals of curves admitting one-parameter groups of infinitesimal point transformations. (*Proc. Nat. Ac. of Sc. Amer.*, Vol. 11, 1925.)
(122) MICHAL (A. B.).—Integro-differential expressions invariant under Volterra's group of transformations (*Ann. of Math.*, Princeton, 2nd Series, Vol. 26, 1925.)

(123) MICHAL (A. B.).—Functional invariants, with continuity of order p, of one-parameter Fredholm and Volterra transformation groups. (*Bull. American Math. Soc.*, Vol. 31, 1925.)
(124) MICHAL (A. B.).—Affinely connected function space manifolds. (*Amer. Journ. Math.*, Vol. I, 1928.)
(125) MOISIL (G. C.).—Sur les variétés fonctionnelles. (*C. R. de l'Acad. des Sciences de Paris*, Vol. 187, 1928.)
(126) MOISIL (G. C.).—Sur les groupes fonctionnelles. (*C. R. de l'Acad. des Sciences de Paris*, Vol. 188, 1929.)
(127) MOISIL (G. C.).—La mécanique analytique des systèmes continus. (*Thèse Fac. Sci.*, Bucarest, 1929.)
(128) MOORE (E. H.).—On a form of general analysis with applications to differential and integral equations. (*Atti. Congr. Int.*, Rome, 1908, Vol. II.)
(129) MOORE (E. H.).—An introduction to a form of general analysis. (*Newhaven Math. Coll.*, Yale Univ. Press, Newhaven, 1910.)
(130) MYLLER (A.).—Randwertaufgaben bei hyperbolischen Differentialgleichungen. (*Math. Ann.*, Vol. LXVIII, 1910.)
(131) MYLLER (A.) and LEBEDEFF (W.).—Über die Anwendung der Integralgleichungen in einer parabolischen Randwertaufgabe. (*Math. Ann.*, Vol. LXVI, 1908.)
(132) NICHOLSON (J. W.).—The electrification of two parallel circular discs. (*Phil. Trans. of Royal Soc. of London*, Series A, Vol. 224, 1924.)
(133) OSEEN (C. W.).—Über die Bedeutung der Integralgleichungen in der Theorie der Bewegung einer reibenden unzusammendrückbaren Flüssigkeit. (*Ark. Mat., Astr., Fys.* (Stockholm), Vol. 6, 1910.)
(134) PEANO (G.).—Definizione dell' area d' una superficie. (*R. Acc. dei Lincei. Rend.*, Vol. 6, 1890.)
(135) PICARD (E.).—Sur quelques applications de l'équation fonctionnelle de M. Fredholm. (*R. Circ. Mat. di Palermo*, Vol. XXII, 1906.)
(136) PICARD (E.).—Sur une équation fonctionnelle se présentant dans la théorie de la distribution de l'électricité avec l'hypothèse de Neumann. (*C. R. de l'Acad. des Sciences de Paris*, Vol. 165; *Ann. de l'École Norm.*, Vol. 34, 1917.)
(137) PICK (G.).—Sur l'évaluation des distances dans l'espace fonctionnel. (*C. R. de l'Acad. des Sciences de Paris*, Vol. 158, 1914.)
(138) PICK (G.).—Distanzschätzungen im Funktionenraum [cf. FRANK, (56)].
(139) PICONE (M.).—Sopra un problema dei valori al contorno nelle equazioni iperboliche alle derivate parziali del second'ordine e sopra una classe di equazioni integrali che a quello si riconnettono. (*Rend. Circ. Mat. di Palermo*, Vol. XXXI, 1911.)
(140) PINCHERLE (S.).—Équations et opérations fonctionnelles. (*Encycl. des Sc. Math.*, Tome II, Vol. V, fasc. 26, 1906.)
(141) PINCHERLE (S.).—Sull'inversione degli integrali definiti. (*Mem. Soc. Ital. delle Scienze*, Vol. XV, 1908.)
(142) PINCHERLE (S.).—Appunti di calcolo funzionale. (*Mem. Acc. di Bologna*, Vol. VIII, Series 6, 1911.)

(143) PINCHERLE (S.).—Sulle operazioni lineari permutabili colla derivazione. (*R. Acc. delle Science dell' Ist. di Bologna*, 1922.)
(144) PINCHERLE (S.) and AMALDI (U.).—Le operazioni distributive e le loro applicazioni all'analisi. (Bologna, Zanichelli, 1901.)
(145) POINCARÉ (H.).—Leçons de mécanique céleste, Vol. III, Chap. X, Théorie des marées. (Paris, Gauthier-Villars, 1910.)
(146) POOLE (E. G. C.).—A new integral equation satisfied by the solutions of a certain linear differential equation which occurs in the theory of electrical oscillations and of the tides. (*Messenger of Math.*, Vol. 49.)
(147) POPOVICI (C.).—Sur les équations intégro-fonctionnelles. (*R. Acc. dei Lincei. Rend.*, Vol. X, Series 6, 1929.)
(148) POYNTING (J. H.).—Transfer of energy in the electromagnetic field. (*London Phil. Trans.*, 1884, Vol. 175.)
(149) PROUDMAN (M.)—Free and forced longitudinal tidal motion in a lake. (*Proc. London Math. Soc.*, 2nd Series, Vol. XV, Part III, 1915.)
(150) RICHARDSON (R. G. D.).—A new method in boundary problems for differential equations. (*Trans. American Math. Soc.*, Vol. XVIII, 1917.)
(151) RISSER (H.).—Sur une application de l'équation de Volterra au problème de la répartition par âge dans les milieux à effectif constant. (*C. R. de l'Acad. des Sciences de Paris*, Vol. 171, 1920.)
(152) Roos (C. J.).—A mathematical theory of competition. (*American Journ. of Math.*, Vol. XLVII, 1925.)
(153) Roos (C. J.).—A dynamical theory of economics. (*Journal of Political Economy*, Vol. 35, 1927.)
(154) Roos C. J.).—Dynamical economics. (*Proc. Nat. Acad. Sci.*, Vol. 13, 1927.)
(155) Roos (C. J.).—A dynamical theory of economic equilibrium. (*Proc. Nat. Acad. Sci.*, Vol. 13, 1927.)
(156) Roos (C. J.).—A mathematical theory of depreciation and replacement. (*Amer. Journ. Math.*, Vol. 50, 1928.)
(157) Roos (C. J.).—Generalized Lagrange problems in the calculus of variations. (*Trans. of American Math. Soc.*, Vol. 30, 1928.)
(158) ROSTER (W.).—Over de Theorie der hysteresis volgens Volterra. (*Anich. Akad. Versl.*, 29, 1920.)
(159) SBRANA (F.).—Sul potenziale di un disco con distribuzione simmetrica. (*R. Acc. dei Lincei. Rend.*, Vol. XXXIII, Series 5, 1924.)
(160) SBRANA (F.).—Di un' equazione integrale, che si presenta nella teoria statistica dell' effetto fotoelettrico. (*R. Acc. dei Lincei. Rend.*, Vol. I, Series 6, 1925.)
(161) SCHÖNBAUM (E.).—Application of Volterra's integral equations to the problems of mathematical statistics. (In Czech.) (*Rozpravy Ceski Akademie*, XXVI, 1917.)
(162) SCHÖNBAUM (E.).—A contribution to the mathematical theory of old-age disability insurance. (In Czech.) *Casopis pro pěstorani mathematiky a fysiky*, XLVII, 1918.)
(163) SCHWARZ (H.).—Sur une définition erronée de l'aire d'une surface courbe. (*Ges. Math. Abhandl.*, Berlin, Vol. II, 1890.)

(164) SERINI (R.).—Sulle leggi ereditarie che conservano i massimi. (Notes I and II.) (*R. Ist. Veneto*, 1917, 1919.)

(165) SIERSPÍNSKI (W.).—La notion de dérivée comme base d'une théorie des ensembles abstraits. (*Math. Ann.*, Vol. 97, 1926.)

(166) STRANEO (P.).—Sull'espressione dei fenomeni ereditari. (*R. Acc. dei Lincei. Rend.*, Vol. I, Series 6, 1925.)

(167) TAMARKIN (J. D.).—On Volterra's integro-functional equation. (*Trans. of American Math. Soc.*, Vol. 28, 1926.) (This paper contains a bibliography of the subject.)

(168) TEDONE (O.).—Su l'inversione di alcuni integrali e la integrazione delle equazioni a derivate parziali col metodo delle caratteristiche. (*R. Acc. dei Lincei. Rend.*, Vol. XXIII, Series 5, 1914.)

(169) TEDONE (O.).—Su alcune altre formole d'inversione collegate col metodo d'integrazione di Riemann. (*R. Acc. dei Lincei. Rend.*, Vol. XXIX, Series 5, 1920.)

(170) TEICHMANN (S.).—Mechanische Probleme, die auf belastete Integralgleichungen führen. (*Diss. Univ. Breslau*, 1918.)

(171) TREFFTZ (E.).—Schwingungsprobleme und Integralgleichungen. (*Math. Ann.*, Vol. 87, 1922.)

(172) URYSOHN (P.).—Un théorème sur la puissance des ensembles ordonnées. (*Fundamenta Math.*, Vol. V-VI, 1924.) (For other papers by this author see (208).)

(173) VESSIOT (E.).—Sur les groupes fonctionnels et les équations intégro-différentielles linéaires. (*C. R. de l'Acad. des Sciences de Paris*, Vol. 154, 1912.)

(174) VESSIOT (E.).—Sur les fonctions permutables et les groupes continus de transformations fonctionnelles linéaires. (*C. R. de l'Acad. des Sciences de Paris*, Vol. 154, 1912.)

(175) VITALI (G.).—Analisi delle funzioni a variazione limitata. (*Rend. Circ. Mat. Palermo*, Vol. 46, 1922.)

(176) VITALI (G.).—Geometria nello spazio hilbertiano. (Bologna, 1929.)

(177) VITERBI (A.).—Sull'operazione funzionale rappresentata da un integrale definito riguardata come elemento d'un calcolo. (*R. Acc. dei Lincei. Rend.*, Vol. VI, Series 5, 1897.)

(178) VITERBI (A.).—Sull'operazione funzionale rappresentata da un integrale definito, considerata come elemento d'un calcolo. (*Ann. di Mat.*, Vol. XXVI, 1897.)

(179) VITERBI (A.).—Sull'operazione funzionale rappresentata da un integrale definito, riguardata come elemento d'un calcolo. (*Ann. di Mat.*, Vol. 3, Series 3, 1909.)

(180) VOLTERRA (V.).—Sul fenomeno delle seiches. (*Nuovo Cimento* (Pisa), Series 4, Vol. VIII, 1898.)

(181) VOLTERRA (V.).—Sulle equazioni integro-differenziali della teoria dell'elasticità. (*R. Acc. dei Lincei. Rend.*, Vol. XVIII, Series 5, 1908.)

(182) VOLTERRA (V.).—Equazioni integro-differenziali dell' elasticità nel caso della isotropia. (*R. Acc. dei Lincei. Rend.*, Vol. XVIII, Series 5, 1909.)

(183) VOLTERRA (V.).—Soluzione delle equazioni integro-differenziali dell' elasticità nel caso di una sfera isotropa. (*R. Acc. dei Lincei. Rend.*, Vol. XIX, Series 5, 1910.)
(184) VOLTERRA (V.).—Deformazioni di una sfera elastica soggetta a date tensioni nel caso ereditario. (*R. Acc. dei Lincei. Rend.*, Vol. XIX, Series 5, 1910.)
(185) VOLTERRA (V.).—Vibrazioni elastiche nel caso della eredità. (*R. Acc. dei Lincei. Rend.*, Vol. XXI, Series 5, 1912.)
(186) VOLTERRA (V.).—Sui fenomeni ereditari. (*R. Acc. dei Lincei. Rend.*, Vol. XXII, Series 5, 1913.)
(187) VOLTERRA (V.).—Variazioni e fluttuazioni del numero d'individui in specie animali conviventi. (*R. Comit. Talass. It.*, Mem. 131, 1927.)
(188) VOLTERRA (V.).—La teoria dei funzionali applicata ai fenomeni ereditari. (*Atti Congr. Int. Mat.*, Bologna, 1928, Vol. I.)
(189) VOLTERRA (V.).—Sur la théorie mathématique des phénomènes héréditaires. (*Journ Math. Pur. Appl.*, Vol. 7, 1928.)
(190) VOLTERRA (V.).—Alcune osservazioni sui fenomeni ereditarii. (*R. Acc. dei Lincei. Rend.*, Vol. IX, Series 6, 1929.)
(191) WIECHERT (J. E.).—Gesetze der elastischen Nachwirkung für konstante Temperatur. (*Wied. Ann.*, Vol. 50.)
(192) WIENER (W.).—The mean of a functional of arbitrary elements. (*Bull. Massachusetts Inst. of Technology*, Series II, No. 1, 1921.)
(193) WIENER (W.).—The average of an analytic functional. (*Proc. N. Academy of Sc.* (U. S. A.), Vol. 7, 1921.)
(194) WIENER (N.).— Limit in terms of continuous transformation. (*Bull. Soc. Math. de France*, Vol. 50, 1922.)
(195) WIENER (N.).—Differential space. (*Bull. Massachusetts Inst. of Technology*, Series II, No. 60, 1923.)
(196.) WIENER (N.).—The average value of a functional. (*Proc. London Math. Soc.*, Series 2, Vol. 22, Part 6, 1924.)
(197) ZAREMBA (S.).—Contribution à la théorie d'une équation fonctionnelle de la physique. (*Rend. Circ. Mat. di Palermo*, Vol. XIX, 1905.)

SUPPLEMENT[1]

(198) BIRTWISTLE (G.).—The new quantum mechanics. (Cambridge, University Press, 1928.)
(199) DE DONDER (T.).—Sur les équations canoniques de Hamilton-Volterra. (*Mémoires publiés par la classe des sciences de l'Acad. Royale de Belgique*, IIe Série, t. III, 1911.)
(200) DONATI (L.).—Sul lavoro di deformazione dei sistemi elastici. (*Memorie della R. Accad. delle Scienze dell' Ist. di Bologna*, Series IV, Vol. IX, 1888.)
(201) GIORGI (G.).—On the functional dependence of physical variables. (*Proc. Int. Math. Cong.*, Toronto, 1924.)

---

[1] These items were added to the Bibliography too late to insert them in their proper places in the preceding list.

(202) GIORGI (G.).—Sulle funzioni delle matrici. (*R. Acc. dei Lincei. Rend.*, Series 6, Vol. VII, 1928.)
(203) GIORGI (G.).—Nuove osservazioni sulle funzioni delle matrici. (*R. Acc. dei Lincei. Rend.*, Series 6, Vol. VIII, 1928.)
(204) FRÉCHET (M.).—Number of dimensions of an abstract set. (*Proc. Int. Math. Cong.*, Toronto, 1924.)
(205) FRÉCHET (M.).—L'expression la plus générale de la "distance" sur une droite. (*Proc. Int. Math. Cong.*, Toronto, 1924.)
(206) FRÉCHET (M.).—Sur une représentation paramétrique intrinsèque de la courbe continue la plus générale. (*Proc. Int. Math. Cong.*, Toronto, 1924.)
(207) FRÉCHET (M.).—L'analyse générale et les espaces abstraits. (*Atti Congr. Int. Mat.*, Bologna, 1928, Vol. I.)
(208) FRÉCHET (M.).—Les espaces abstraits et leur théorie considérée comme introduction à l'Analyse générale. (Paris, Gauthier-Villars, 1928.)([1])
(209) HEAVISIDE (O.).—Electrical Papers. (2 Vols., 1872.)
(210) HEAVISIDE (O.).—Electromagnetic Theory. (3 Vols. London, Benn; second reprint 1925.)
(211) KORN (A.).—Anwendung der Heaviside'schen Methode zur Integration der Wärmeleitungsgleichung. (*Sitzungsb. d. Berl. Math. Gesellschaft*, XXVIII. Jahrgang, 1929.)
(212) KORN (A.).—Verallgemeinerung von Herationen für nicht ganze Iterationszahlen. (*Sitzungsb. d. Berl. Math. Gesellschaft*, XXVIII. Jahrgang, 1929.)
(213) LEIBNIZ (G. G.).—Miscellanea Berolinensia, Vol. I (1710); Opera Omnia, Vol. III (Geneva, 1768.)
(214) MORERA (G.).—Un teorema fondamentale nella teorica delle funzioni di una variabile complessa. (*Rend. R. Ist. Lombardo*, Milan, Series II, Vol. XIX, 1886.)
(215) OSEEN (C. W.).—Über einen Satz von Green und über die Definition von Rot. und Div. (*Rend. Circ. Mat. di Palermo*, Vol. 38, 1914.)
(216) OSEEN (C. W.).—Hydrodynamik. (Leipzig, 1927.)
(217) PINCHERLE (S.).—Sulle operazioni funzionali lineari. (*Proc. Int. Math. Cong.*, Toronto, 1924.)
(218) STEINHAUS (H.).—Anwendungen der Funktionalanalysis auf einige Fragen der reellen Funktionentheorie. (*Studia Mathematica* (Lwow), Vol. I, 1929.)
(219) STRANEO (S. L.).—Sulla risoluzione funzionale dei problemi lineari di propagazione del calore. (*R. Acc. dei Lincei. Rend.*, Series 6, Vol. IX, 1929.)
(220) TARDY (P.).—Intorno ad una formula del Leibniz. (*Bull. di bibl. e storia delle Scienze mat. e fis. di Boncompagni*, Vol. I, 1868.)

([1]) Only a concise bibliography of abstract spaces has been given here. A more complete bibliography is to be found in this book by FRÉCHET.

INDEX OF AUTHORS CITED

(The numbers refer to the pages)

Abel 55, 56, 69.
Adams 176, 208.
Alexandroff 205, 208.
Amoroso 167, 206, 208.
Andreoli 134, 208.
Arzelà 13, 35, 173.
Bär 208.
Baeri 167.
Baire 8, 14, 35.
Banach viii, 208.
Barnet 167.
Barnett 208.
Bateman 208.
Bedarida 167.
Bennett 69.
Bernstein 69.
Bertrand 208.
Bianchi 103.
Birkhoff 208.
Birtwistle 200, 217.
Bliss 206, 208.
Block 69.
Bôcher vi, 50, 69, 181.
Bohannan 97.
Boltzmann 208.
Bompiani 135.
Bonferroni 208.
Born 208.
Bouligand 35, 181, 208.
Bourlet 201, 208.
Bray 180, 181, 182, 208.
Browne 69.
Bucht 69.
Burgatti 69.

Calò 200, 209.
Castigliano 187.

Cauchy 74, 87, 95, 103, 131, 183, 192.
Chrystal 209.
Courant vii 24, 35, 209.
Cramer 44, 48, 49.
Crudeli 167.
D'Alembert 149.
Daniele 35, 135, 167.
Daniell 34, 35.
Davis 69, 174, 209.
De Donder 97, 217.
Delsarte vi, 185, 186, 209.
De Toledo v, ix
Dienes 35.
Dines 209.
Dirichlet 4, 46, 47, 56, 57, 165, 173, 174, 175, 181, 182.
Dixon 69.
Donati 187, 217.

Euler 31, 127, 160.
Evans v, vi, vii, viii 35, 47, 58, 68, 69, 70, 101, 134, 135, 154, 155, 167. 168, 180, 181, 182, 206, 208, 209.

Fabri 6, 36, 79.
Fantappiè, v, vi, ix, 53, 70, 184, 200, 201, 202, 203, 204, 209, 210.
Fischer 36, 168.
Forel 177, 210.
Fourier 154, 176.
Frank 210.
Fréchet vi, 8, 9, 10, 13, 15, 20, 21, 27, 29, 36, 97, 156, 168, 202, 204, 205, 210, 211, 218.
Freda viii, 160, 161, 168.
Fredholm 49, 62, 70, 175, 176, 184, 185, 186, 204, 211.
Fubini 168.

220 INDEX OF AUTHORS CITED

Gateaux vii, 12, 33, 36, 37, 164, 187.
Gauss 89, 173.
Giorgi 200, 211, 217, 218.
Goursat 50, 51, 52, 70, 176.
Graffi 149, 171, 196.
Gramegna 168.
Graustein 97.
Green vii, 89, 152, 153, 165, 166, 174, 179, 181.

Hadamard vii, 13, 15, 16, 37, 55, 70, 103, 105, 124, 133, 135, 166, 168, 173, 183, 184, 202, 203, 211.
Hamilton 156, 163, 164, 172.
Heaviside 200, 218.
Hebroni 168.
Hecke 211.
Heisenberg 164, 173, 208, 211.
Helly 37.
Hertz 194. 211.
Hilb 168.
Hilbert vii, 50, 51, 70, 173, 175, 176, 203, 204, 212.
Hill 168.
Hildebrand 37.
Hirahawa 212.
Holmgren 55, 70.
Hooke 193, 194.
Horn 70.
Hostinsky 166, 212.
Hotelling 206, 212.
Hu (Minfu Tat) 168.
Huygens 183, 184.

Ince 70.

Jacobi 155, 156, 163.
Jordan 164, 165, 168, 176.
Joule 199.

Kakeya 37, 70.
Kellogg 208, 212.
Korn 212, 218.
Kostitzin 195, 212.
Kowalewski 183, 184, 212.
Krall 166, 168.

Lagrange 74, 179, 185, 191, 200.
Lalesco 50, 61, 70, 176.
Laplace 86, 88, 89, 90, 92, 150, 153, 164, 174, 181.

Lauricella 53, 70, 135, 212.
Lebesgue 9, 37, 132, 135.
Lefschetz 85, 97.
Leibniz 200, 218.
Lense 212.
Le Roux 70.
Le Stourgeon 37.
Levi 37, 213.
Levi-Civita 70, 213, 184.
Lévy (P.) vi, vii, 11, 12, 18, 37, 70, 71, 159, 166, 169, 205, 213.
Lichtenstein 213.
Lie 184.
Liouville 55, 56, 71.
Lipschitz 140.
Lüsis 135.

Mandelbrojt 56, 71, 95, 97.
Manneback 213.
Marcolongo 213.
Maria 180, 213.
Matteuzzi 180, 213.
Mérian 177, 213.
Michal vi, 184, 213, 214.
Miles 182.
Moisil vi, 185, 214.
Molinari 71.
Moore vi, 204, 214.
Morera 95, 218.
Myller 214.

Nalli 38, 136.
Neumann 175, 178, 182.
Nicholson 214.
Nicolesco 74, 97.
Nörlund 136.

Oseen 214, 218.

Pascal (E.), 38, 97.
Pascal (M.), 97.
Pauli 164, 165, 168, 173, 176, 211.
Peano 173, 214.
Pérès vi, viii, 101, 106, 110, 111, 112, 113, 114, 115, 116, 117, 118, 121, 122, 123, 126, 129, 130, 131, 136, 169, 182.
Picard 42, 53, 71, 139, 214.
Pick 214.
Picone 38, 71, 214.

INDEX OF AUTHORS CITED 221

Pincherle 200, 201, 202, 214, 215, 218.
Poincarè 175, 180, 215.
Poisson 164, 181.
Poli 71.
Pomey 141, 169.
Poole 215.
Popovici 176, 215.
Poynting 198, 215.
Prasad 71.
Proudman 179, 215.

Radon 38.
Rasor 38.
Recchia 136.
Ricci 176.
Richardson 215.
Riemann 55, 71, 74, 86, 173.
Riesz 17, 38.
Risser 206, 215.
Roos 206, 215.
Roster 215.

Sabbatini 66, 71.
Sassmannshausen 170.
Sbrana 71, 215.
Scatizzi 55, 72.
Schlesinger 141, 170.
Schmidt 51, 65, 72, 176.
Schönbaum 170, 206, 215.
Schrödinger 165.
Schwarz 173, 215.
Serini 195, 216.
Servois 200.
Severini 136.
Sierspinski 205, 216.
Sinigallia 136, 170.
Somigliana 166.
Sonine 56, 72.

Stäckel 164.
Steinhaus viii, 15, 38, 218.
Stieltjes 9, 17, 18, 34, 182.
Stokes 158, 159.
Straneo (P.) 189, 195, 216.
Straneo (S. L.) 218.

Tamarkin 176, 216.
Tardy 200, 218.
Taylor 25, 26, 96, 138, 139, 148, 154, 194, 202.
Tedone 68, 72, 216.
Teichmann 216.
Thomson 173.
Tonelli vii, 14, 38, 173.
Trefftz 216.
Tricomi 38, 170.

Urysohn 205, 216.
Usai 72.

Vacca viii.
Vergerio 72, 170.
Vessiot 38, 118, 136, 183, 184, 216.
Vitali 176, 180, 216.
Viterbi 72, 216.
Vivanti 50, 72.
Weierstrass 8, 13, 20, 21, 39, 74, 173.
Wiechert 217.
Wiener 34, 39, 217.
Winter 39, 201.
Whittaker 68. 73.
Wishnewsky 39.

Zanoni 166, 170.
Zaremba 217.
Zeilon 73, 170.

INDEX OF SUBJECTS
(The numbers refer to the pages)

Page

Addition theorems
Huygens' principle in Hadamard's sense 183
Integral addition theorems 105, 133
Integral addition theorems connected with the equations of mathematical physics 183
Bôcher's type of integro-differential equations 181
Calculus of variations as a chapter of theory of functionals 13, 14, 172
Composition
Composition of the first kind. 99
Composition of the second kind 100
Continuity of functionals . . . . 9—14
Derivatives
Derivative by composition 129
Derivatives of functions of lines with respect to the coordinate planes 80
Derivatives with any index whatever 55
Differentiation, derivation, variation of functionals . . 22—25
Necessary and sufficient condition that a functional may be considered as the functional derivative of another functional 157—159
Symmetry of derivatives of functionals. 24
Dirichlet's problem
Evans' method in the case of discontinuous boundary value problems . . . . .180—181
Method of the calculus of variations and functions of lines . . 173
Method of integral equations . . . . . . .174—175
Euler's theorem on homogeneous functions (Extension of) . . . 160
Evans' boundary value problems for Poisson's equation (see Dirichlet's problem) 181
Fantappiè's functional indicatrix . . . . . . .201—203
Frechet's abstract aggregates, abstract spaces . . 8—14, 204—205
Freda's functional derivative equations 160
Fuchs' theory of linear differential equations (Extension of) . . 141
Functionals
Analytic functionals 201
Extremes of a functional 13
Functional derivative equations 155
Functional dynamics 184
Functional fields 7

INDEX OF SUBJECTS

	Page
Functionals in quantum mechanics	203
Functionals in the theory of elasticity (see also Hereditary questions)	187
Functionals of degree n	19
Functionals of the first degree	14
Functional operators	8, 200
Functional power series	21
Functional rotations	185
Functional transformations	183
Functional derivative equations in the theory of isogeneity	90—95

Functional derivative equations of the first order

Canonical integro-differential equations and connected functional derivative equation	163
Functional derivative equations in functions of lines (see Jacobi-Hamilton theory)	82, 88, 155, 156
Functional derivative equations of the first order and connected integro-differential equations	161

Functional derivative equations of the second order

Equations in quantum electrodynamics	165
Equations of Laplace type	164

Functional equations

Branch elements of solutions of functional equations	65—66
General functional equations and implicit functionals	63—65
Generalisations of systems of n equations in n unknowns	40

Functions

Conjugate functions in space	87—90
Functions by composition	129, 132
Functions of hyperspaces	6, 85
Functions of lines	5, 9, 74
Functions of lines of the first degree	79
Isogenic functions	91—95
Order of a function	110

Functions of an infinite number of variables and problems with an infinite number of unknowns 2, 3, 22, 23, 40, 43, 49, 138, 158, 161, 163, 176, 179—180, 184—185, 203

Green's function in integro-differential equations (Extension of) (see Hadamard's differential equation for Green's function) . . 153

Groups (Functional)

Groups in functional kinematics	185
Groups of linear functional transformations	183—184
Group of the functions permutable with a given function	110
Invariant for groups of functional transformations	184

Hadamard's differential equation for Green's function . . . 166

Expression for linear functionals (see Functionals of the first degree) 15

Hereditary questions

Energy equations in hereditary phenomena	196—200
General laws of heredity	188—190
Hereditary action in elastic torsion	147—149, 189

INDEX OF SUBJECTS

	Page
Hereditary dynamics	191
Hereditary elasticity	192
Hereditary electromagnetism	194
Hereditary phenomena	147
Principle of the closed cycle	189
Principle of the dissipation of hereditary action	188

Integral equations
- Applications of the theory of composition to the solution of integral equations 106, 143, 145
- Fredholm's integral equations 41
- Fredholm's linear integral equations of the first kind . . . 51
- Fredholm's linear integral equations of the second kind . . 48
- Integral equations obtained from problems of the calculus of variations 30, 141
- Inversion of multiple integrals 63
- Systems of integral equations 62
- Volterra's integral equations 42
- Volterra's linear integral equations of the first kind . . . 53
- Volterra's linear integral equations of the second kind . . . 43

Integration of functionals 32

Integro-differential equations
- Equations of canonical type (see Functional derivative equations of the first order) 163
- Generalisation of systems of n differential equations with n unknown functions 138
- Integro-differential equations for the determination of groups of permutable functions 112, 142
- Integro-differential equations obtained from problems of the calculus of variations 31, 142
- Integro-differential equations obtained from the theory of permutable functions of the first kind . . . . . . 143—145
- Integro-differential equations obtained from the theory of permutable functions of the second kind . . . . 145—147
- Integro-differential equations for biological species living together 206
- Integro-differential equations of elliptic, hyperbolic, and parabolic type . . . . 149—155
- Systems of integro-differential equations 145

Jacobi-Hamilton theory (Extension of the) 156

Kernels
- Case in which the kernel vanishes 60
- Evans' method by calculation of solvent kernels . . . 67
- Evans' periodic kernels 134
- Kernels of the closed cycle 189
- Logarithmic kernels 58
- Singular kernels 54, 58
- Solvent kernels 44, 67
- Symmetrical kernels 50

Logarithms by composition 124

INDEX OF SUBJECTS

	Page
Pérès' transformations	.114—118

Permutability
 Determination of the functions permutable with a given function
  (see Groups [Functional]) . . . . . . .110—114
 Functions permutable of the first kind with unity . .108—110
 Permutability of the first and of the second kind 100

Poisson's parentheses (Extension of). 164

Polydromy of function of lines . . . . 82—85

Powers
 Powers and polynomials by composition 101
 Powers by composition with a fractional exponent . . . 118
 Powers by composition with a negative exponent . . .122—124

Seiches (Oscillations of lakes) 177, 179

Series
 Series by composition of the first kind 102
 Series by composition of the second kind 132

Stäckel's theorem of the separation of the variables (Extension of) 164

Stokes' theorem (Extension of) 158

Taylor's theorem for functionals (Extension of) 25

Vibrations of membranes 178

Weierstrass' theorem on continuous functions represented as limits
 of polynomials (Extension of) 20

A series of hardcover reprints of major works in mathematics, science and engineering. All editions are 5⅜ × 8½ unless otherwise noted.

Mathematics

Theory of Approximation, N. I Achieser. Unabridged republication of the 1956 edition. 320pp. 49543-4

The Origins of the Infinitesimal Calculus, Margaret E. Baron. Unabridged republication of the 1969 edition. 320pp. 49544-2

A Treatise on the Calculus of Finite Differences, George Boole. Unabridged republication of the 2nd and last revised edition. 352pp. 49523-X

Space and Time, Emile Borel. Unabridged republication of the 1926 edition. 15 figures. 256pp. 49545-0

An Elementary Treatise on Fourier's Series, William Elwood Byerly. Unabridged republication of the 1893 edition. 304pp. 49546-9

Substance and Function & Einstein's Theory of Relativity, Ernst Cassirer. Unabridged republication of the 1923 double volume. 480pp. 49547-7

A History of Geometrical Methods, Julian Lowell Coolidge. Unabridged republication of the 1940 first edition. 13 figures. 480pp. 49524-8

Linear Groups with an Exposition of Galois Field Theory, Leonard Eugene Dickson. Unabridged republication of the 1901 edition. 336pp. 49548-5

Continuous Groups of Transformations, Luther Pfahler Eisenhart. Unabridged republication of the 1933 first edition. 320pp. 49525-6

Transcendental and Algebraic Numbers, A. O. Gelfond. Unabridged republication of the 1960 edition. 208pp. 49526-4

Lectures on Cauchy's Problem in Linear Partial Differential Equations, Jacques Hadamard. Unabridged reprint of the 1923 edition. 320pp. 49549-3

The Theory of Branching Processes, Theodore E. Harris. Unabridged, corrected republication of the 1963 edition. xiv+230pp. 49508-6

The Continuum, Edward V. Huntington. Unabridged republication of the 1917 edition. 4 figures. 96pp. 49550-7

Lectures on Ordinary Differential Equations, Witold Hurewicz. Unabridged republication of the 1958 edition. xvii+122pp. 49510-8

Mathematical Methods and Theory in Games, Programming, and Economics: Two Volumes Bound as One, Samuel Karlin. Unabridged republication of the 1959 edition. 848pp. 49527-2

Famous Problems of Elementary Geometry, Felix Klein. Unabridged reprint of the 1930 second edition, revised and enlarged. 112pp. 49551-5

Lectures on the Icosahedron, Felix Klein. Unabridged republication of the 2nd revised edition, 1913. 304pp. 49528-0

On Riemann's Theory of Algebraic Functions, Felix Klein. Unabridged republication of the 1893 edition. 43 figures. 96pp. 49552-3

A Treatise on the Theory of Determinants, Thomas Muir. Unabridged republication of the revised 1933 edition. 784pp. 49553-1

A Survey of Minimal Surfaces, Robert Osserman. Corrected and enlarged republication of the work first published in 1969. 224pp. 49514-0

The Variational Theory of Geodesics, M. M. Postnikov. Unabridged republication of the 1967 edition. 208pp. 49529-9

DOVER PHOENIX EDITIONS

An Introduction to the Approximation of Functions, Theodore J. Rivlin. Unabridged republication of the 1969 edition. 160pp. 49554-X
An Essay on the Foundations of Geometry, Bertrand Russell. Unabridged republication of the 1897 edition. 224pp. 49555-8
Elements of Number Theory, I. M. Vinogradov. Unabridged republication of the first edition, 1954. 240pp. 49530-2
Asymptotic Expansions for Ordinary Differential Equations, Wolfgang Wasow. Unabridged republication of the 1976 corrected, slightly enlarged reprint of the original 1965 edition. 384pp. 49518-3

Physics

Semiconductor Statistics, J. S. Blakemore. Unabridged, corrected, and slightly enlarged republication of the 1962 edition. 141 illustrations. xviii+318pp. 49502-7
Wave Propagation in Periodic Structures, L. Brillouin. Unabridged republication of the 1946 edition. 131 illustrations. 272pp. 49556-6
The Conceptual Foundations of the Statistical Approach in Mechanics, Paul and Tatiana Ehrenfest. Unabridged republication of the 1959 edition. 128pp. 49504-3
The Analytical Theory of Heat, Joseph Fourier. Unabridged republication of the 1878 edition. 20 figures. 496pp. 49531-0
States of Matter, David L. Goodstein. Unabridged republication of the 1975 edition. 154 figures. 4 tables. 512pp. 49506-X
The Principles of Mechanics, Heinrich Hertz. Unabridged republication of the 1900 edition. 320pp. 49557-4
Thermodynamics of Small Systems, Terrell L. Hill. Unabridged and corrected republication in one volume of the two-volume edition published in 1963–1964. 32 illustrations. 408pp. 6½ x 9¼. 49509-4
Theoretical Physics, A. S. Kompaneyets. Unabridged republication of the 1961 edition. 56 figures. 592pp. 49532-9
Quantum Mechanics, H. A. Kramers. Unabridged republication of the 1957 edition. 14 figures. 512pp. 49533-7
The Theory of Electrons, H. A. Lorentz. Unabridged reproduction of the 1915 edition. 9 figures. 352pp. 49558-2
The Principles of Physical Optics, Ernst Mach. Unabridged republication of the 1926 edition. 279 figures. 10 portraits. 336pp. 49559-0
The Scientific Papers of James Clerk Maxwell, James Clerk Maxwell. Unabridged republication of the 1890 edition. 197 figures. 39 tables. Total of 1,456pp.
Volume I (640pp.) 49560-4; *Volume II* (816pp.) 49561-2
Vectors and Tensors in Crystallography, Donald E. Sands. Unabridged and corrected republication of the 1982 edition. xviii+228pp. 49516-7
Principles of Mechanics and Dynamics, Sir William (Lord Kelvin) Thompson and Peter Guthrie Tait. Unabridged republication of the 1912 edition. 168 diagrams. Total of 1,088pp. *Volume I* (528pp.) 49562-0; *Volume II* (560pp.) 49563-9
Treatise on Irreversible and Statistical Thermophysics: An Introduction to Nonclassical Thermodynamics, Wolfgang Yourgrau, Alwyn van der Merwe, and Gough Raw. Unabridged, corrected republication of the 1966 edition. xx+268pp. 49519-1

Engineering

Principles of Aeroelasticity, Raymond L. Bisplinghoff and Holt Ashley. Unabridged, corrected republication of the original 1962 edition. xi+527pp. 49500-0
Statics of Deformable Solids, Raymond L. Bisplinghoff, James W. Mar, and Theodore H. H. Pian. Unabridged and corrected Dover republication of the edition published in 1965. 376 illustrations. xii+322pp. 6½ x 9¼. 49501-9